An Introduction to Air Pollution Control Engineering

2nd Edition

J. Paul Guyer, P.E., R.A.

Editor

The Clubhouse Press
El Macero, California

CONTENTS

(This publication is adapted from the *Unified Facilities Criteria* and other resources of the United States government which are in the public domain, have been authorized for unlimited distribution, and are not copyrighted.)

CHAPTER 1
CYCLONE COLLECTORS

1.1 CYCLONE. The cyclone is a widely used type of particulate collection device in which dust-laden gas enters tangentially into a cylindrical or conical chamber and leaves through a central opening. The resulting vortex motion or spiraling gas flow pattern creates a strong centrifugal force field in which dust particles, by virtue of their inertia, separate from the carrier gas stream. They then migrate along the cyclone walls by gas flow and gravity and fall into a storage receiver. In a boiler or incinerator installation this particulate is composed of fly-ash and unburned combustibles such as wood char. Two widely used cyclones are illustrated in figure 6-I.

1.2 CYCLONE TYPES.

1.2.1 CYCLONES ARE GENERALLY classified according to their gas inlet design and dust discharge design, their gas handling capacity and collection efficiency, and their arrangement. Figure 6-2 illustrates the various types of gas flow and dust discharge configurations employed in cyclone units. Cyclone classification is illustrated in table 6-I.

1.2.2 CONVENTIONAL CYCLONE. The most commonly used cyclone is the medium efficiency, high gas throughput (conventional) cyclone. Typical dimensions are illustrated in figure 6-3. Cyclones of this type are used primarily to collect coarse particles when collection efficiency and space requirements are not a major consideration. Collection efficiency for conventional cyclones on 10 micron particles is generally 50 to 80 percent.

1.2.3 HIGH EFFICIENCY CYCLONE. When high collection efficiency (80-95 percent) is a primary consideration in cyclone selection, the high efficiency single cyclone is commonly used (See figure 6-4). A unit of this type is usually smaller in diameter than the conventional cyclone, providing a greater separating force for the same inlet velocity and a shorter distance for the particle to migrate before reaching the cyclone walls. These units may be used singly or arranged in parallel or series as shown in figure 6-5. When arranged in parallel they have the advantage of handling larger gas volumes at increased efficiency for the same power consumption of a conventional unit. In parallel they also have the ability to reduce headroom space requirements below that of a single cyclone handling the same gas volumes by varying the number of units in operation.

1.2.4 MULTICYCLONES. When very large gas volumes must be handled and high collection efficiencies are needed a multiple of small diameter cyclones are usually nested together to form a multicyclone. A unit of this type consists of a large number of elements joined together with a common inlet plenum, a common outlet plenum, and a common dust hopper. The multicyclone elements are usually characterized by having a small diameter and having axial type inlet vanes. Their performance may be hampered by poor gas distribution to each element, fouling of the small diameter dust outlet, and air leakage or back flow from the dust bin into the cyclones. These problems are offset by the advantage of the multicyclone's increased collection efficiency over the single high efficiency cyclone unit. Problems can be reduced with proper plenum and dust discharge design. A typical fractional efficiency curve for multicyclones is illustrated in figure 6-6.

1.2.5 WET OR IRRIGATED CYCLONE. Cyclones may be operated wet in order to improve efficiency and prevent wall buildup or fouling (See fig. 6-7). Efficiency is higher for this type of operation because dust particles, once separated, are trapped in a liquid film on the cyclone walls and are not easily re-entrained. Water is usually sprayed at the rate of 5 to 15 gallons per 1,000 cubic feet (ft3) of gas. Wet operation has the additional advantages of reducing cyclone erosion and allowing the hopper to be placed remote from the cyclones. If acids or corrosive gases are handled, wet operation may result in increased corrosion. In this case, a corrosion resistant lining may be needed. Re-entrainment caused by high values of tangential wall velocity or accumulation of liquid at the dust outlet can occur in wet operation. However, this problem can be eliminated by proper cyclone operation. Wet operation is not currently a common procedure for boilers and incinerators.

1.3 CYCLONE COLLECTION EFFICIENCY. The ability of a cyclone to separate and collect particles is dependent upon the particular cyclone design, the properties of the gas and the dust particles, the amount of dust contained in the gas, and the size distribution of the particles. Most efficiency determinations are made in tests on a geo-'metrically similar prototype of a specific cyclone design in which all of the above variables are accurately known. When a particular design is chosen it is usually accurate to estimate cyclone collection efficiency based upon the cyclone manufacturer's efficiency curves for handling a similar dust and gas. All other methods of determining cyclone efficiency are estimates and should be treated as such.

1.3.1 PREDICTING CYCLONE COLLECTION EFFICIENCY. A particle size distribution curve for flue gas entering a cyclone is used in conjunction with a cyclone fractional efficiency curve in order to determine overall cyclone collection efficiency.

1.3.1.1 A PARTICLE SIZE DISTRIBUTION CURVE shows the weight of the particles for a given size range in a dust sample as a percent of the total weight of the sample. Particle size distributions are determined by gas sampling and generally conform to statistical distributions. See figure 6-8.

1.3.1.2 A FRACTIONAL CYCLONE EFFICIENCY curve is used to estimate what weight percentage of the particles in a certain size range will be collected at a specific inlet gas flow rate and cyclone pressure drop. A fractional efficiency curve is best determined by actual cyclone testing and may be obtained from the cyclone manufacturer. A typical manufacturer's fraction efficiency curve is shown on figure 6-9.

1.3.1.3 CYCLONE COLLECTION EFFICIENCY is determined by multiplying the percentage weight of particles in each size range (size distribution curve) by the collection efficiency corresponding to that size range (fractional efficiency curve), and adding all weight collected as a percentage of the total weight of dust entering the cyclone.

© J. Paul Guyer   2021   4

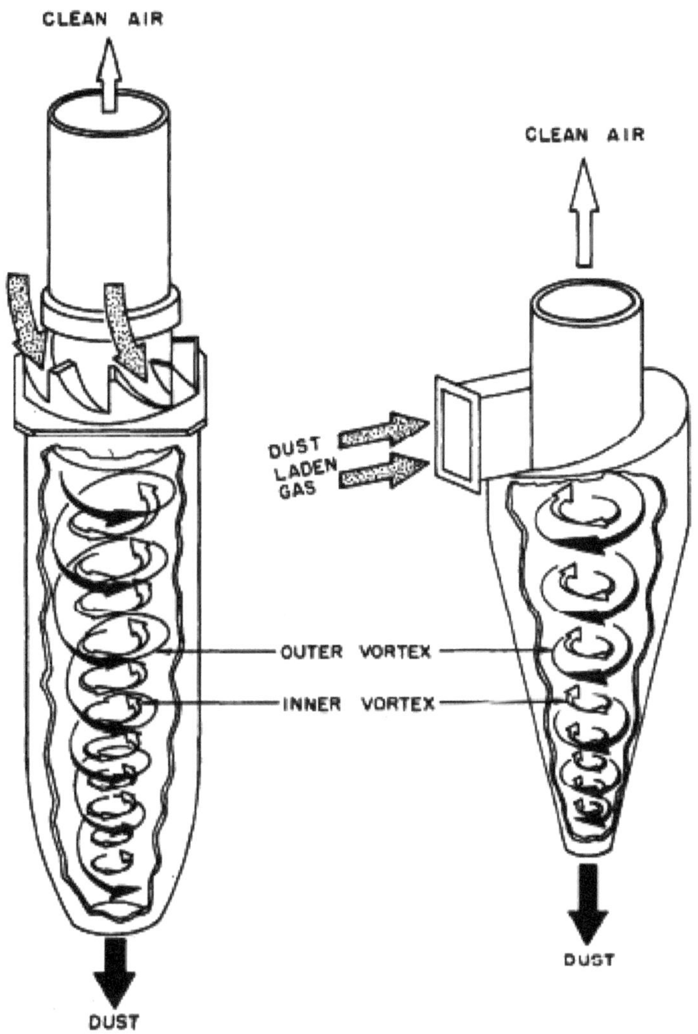

Figure 6-I

Cyclone configuration

© J. Paul Guyer   2021

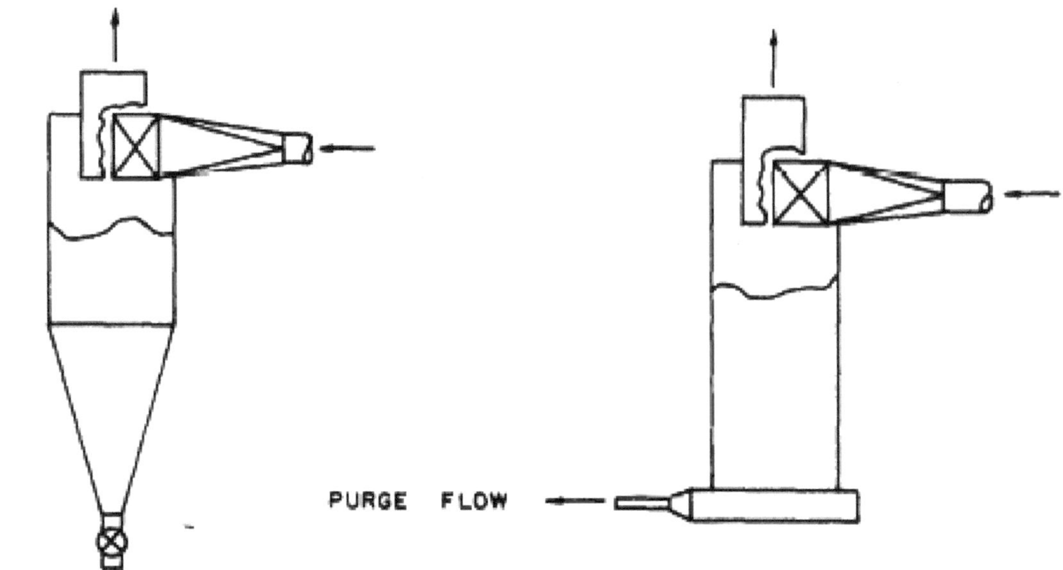

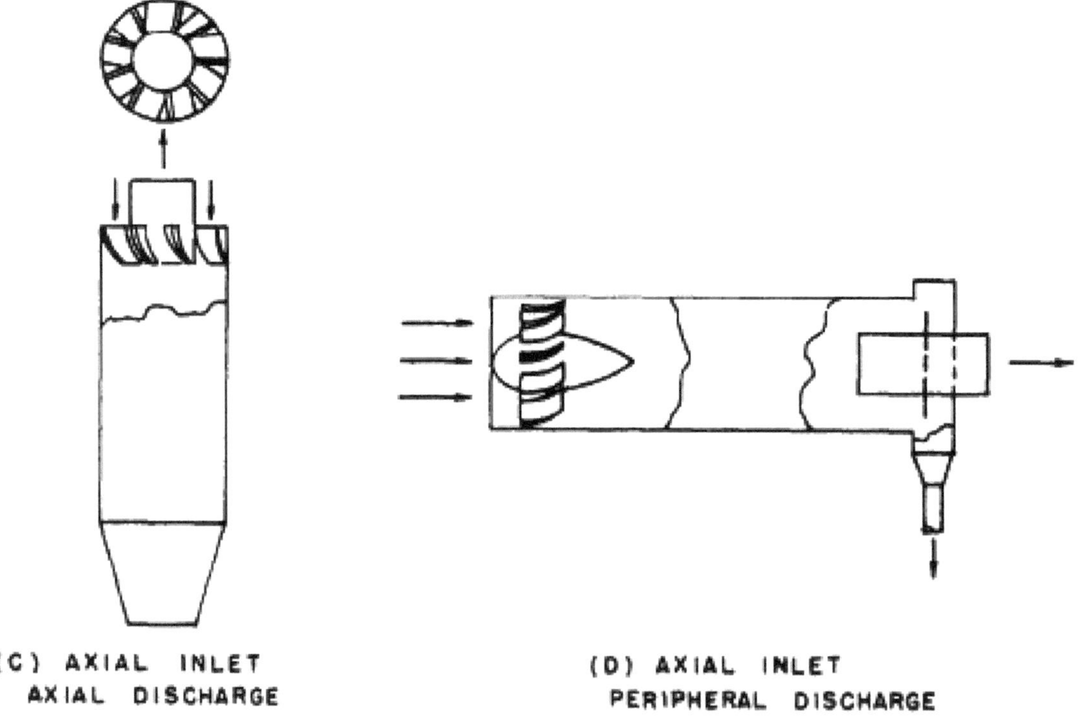

Figure 6-2

Types of cyclones in common use

Type	Body Diameter (ft.)	Gas Flow (ft^3/) min	Pressure Drop (in H_2O)	Inlet Velocity ft/s	Collection Efficiency (%)	Application	Other
Conventional	4-12	1,000-20,000	.5-2	20-70	50-80	Material handling Exhaust gas precleaner	Large headroom requirements. Limited to large coarse particals, large grain loadings.
High efficiency single cyclone	less than 3	100-2,000	2-6	50-70	80-95	Industrial boiler particulate control	Smaller space requirement. Parallel arrangement, inlet vane flow controls needed continuous dust removal system purge operation.
Multicyclones	.5-1	30,000 and up	3-6	50-70	90-95	Industrial and utility boiler. Particulate control	Plenums required. Problems: gas recirculation fouling. Continuous dust removal system, flow control.
Irrigated cyclone (wet) high efficiency single unit	less than 3	100-2,000	2-6	50	90-95	Boiler application (low sulfur fuel) (low gas temperature).	Water rate 5-15 gal/1,000 ft^3/min. Corrosion resistant materials.

NOTE: Cyclone collection efficiency must be evaluated for each specific application due to the sensitivity of cyclone performance on gas and dust properties and loadings. Refer to Table 6-8 for expected collection efficiency estimates.

Table 6-1

Cyclone classification

© J. Paul Guyer 2021

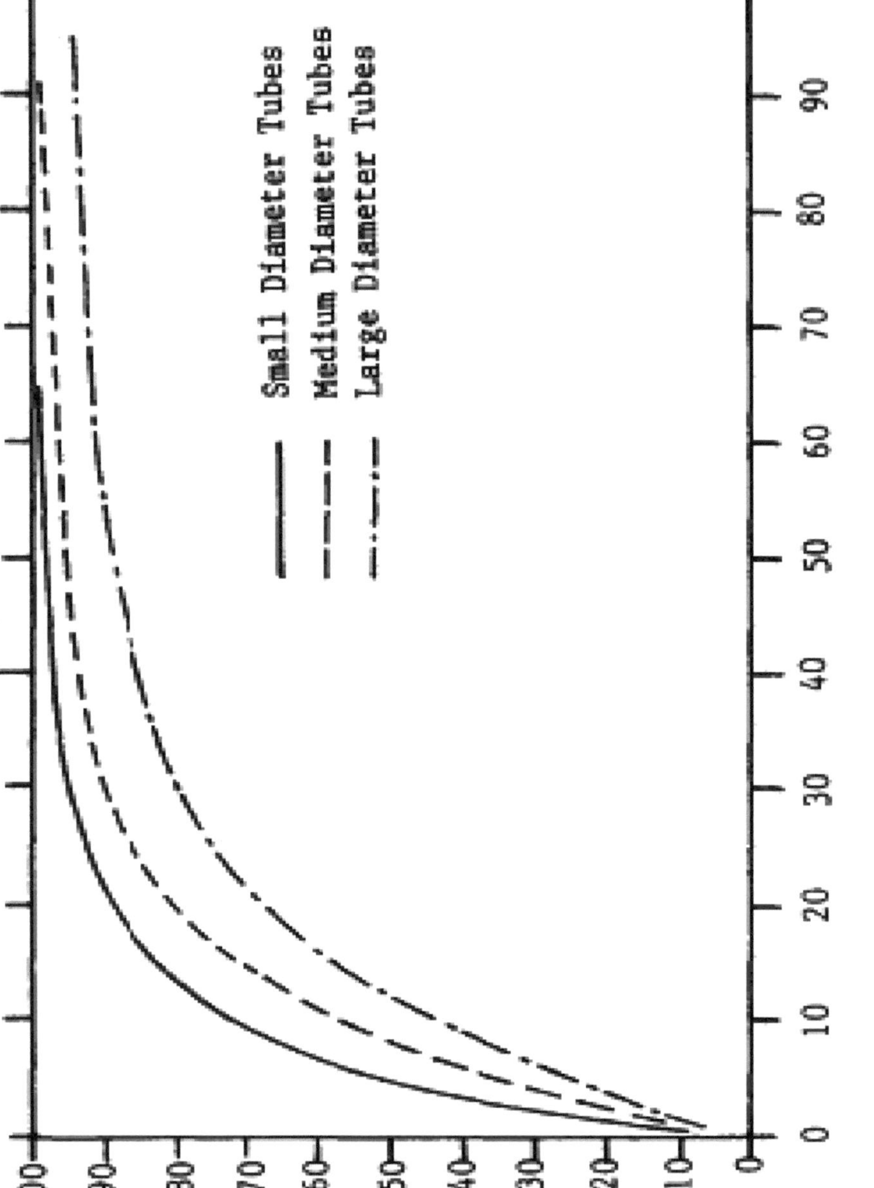

Figure 6-3

Relative effect of cyclone dimensions on efficiency.

© J. Paul Guyer   2021

1.4 CYCLONE PRESSURE DROP AND ENERGY REQUIREMENTS

1.4.1 PRESSURE DROP. Through any given cyclone there will be a loss in static pressure of the gas between the inlet ductwork and the outlet ductwork. This pressure drop is a result of entrance and exit losses, frictional losses and loss of rotational kinetic energy in the exiting gas stream. Cyclone pressure drop will increase as the square of the inlet velocity.

1.4.2 CYCLONE ENERGY REQUIREMENTS. Energy requirements in the form of fan horsepower are directly proportional to the volume of gas handled and the cyclone resistance to gas flow. Fan energy requirements are estimated at one quarter horsepower per 1000 cubic feet per minute (cfm) of actual gas volume per one inch, water gauge, pressure drop. Since cyclone pressure drop is a function of gas inlet and outlet areas, cyclone energy requirements (for the same gas volume and design collection efficiency) can be minimized by reducing the size of the cyclone while maintaining the same dimension ratios. This means adding more units in parallel to handle the required gas volume. The effect on theoretical cyclone efficiency of using more units in parallel for a given gas volume and system pressure drop is shown in figure 6-10. The increased collection efficiency gained by compounding cyclones in parallel can be lost if gas recirculation among individual units is allowed to occur.

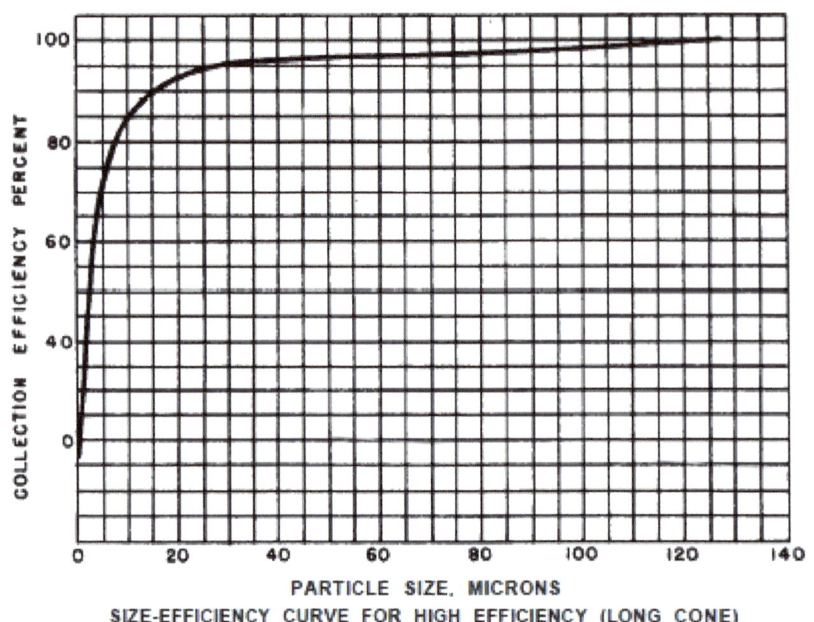

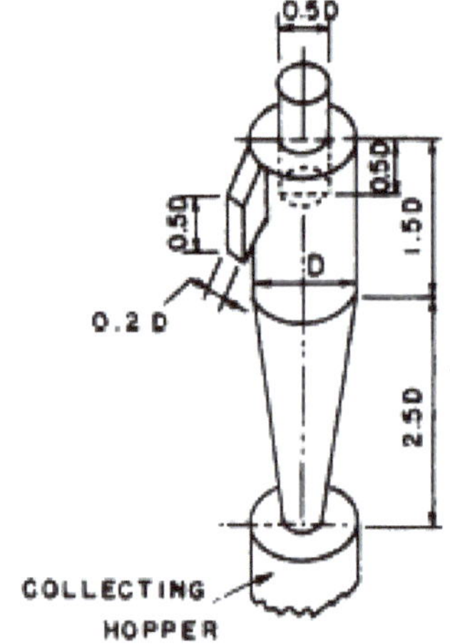

Figure 6-4

Efficiency curve and dimensions for a high efficiency single cyclone

1.5 APPLICATION

1.5.1 PARTICULATE COLLECTION. Cyclones are used as particulate collection devices when the particulate dust is coarse, when dust concentrations are greater than 3 grains per cubic foot (gr/ft3), and when collection efficiency is not a critical requirement. Because collection efficiencies are low compared to other collection equipment, cyclones are often used as pre-cleaners for other equipment or as a final cleaner to improve overall efficiency.

1.5.2 PRE-CLEANER. Cyclones are primarily used as precleaners in solid fuel combustion systems such as stoker fired coal burning boilers where large coarse particles may be generated. The most common application is to install a cyclone ahead of an electrostatic precipitator. An installation of this type is particularly efficient because the cyclone exhibits an increased collection efficiency during high gas flow and dust loading conditions, while the precipitator shows and increase in collection efficiency during decreased gas flow and dust loading. The characteristics of each type of equipment compensate for the other, maintaining good efficiency over a wide range of operating flows and dust loads. Cyclones are also used as pre-cleaners when large dust loads and coarse abrasive particles may affect the performance of a secondary collector. They can also be used for collection of unburned particulate for re-injection into the furnace.

1.5.3 FINE PARTICLES. Where particularly fine sticky dust must be collected, cyclones more than 4 to 5 feet in diameter do not perform well. The use of small diameter multicyclones produces better results but may be subject to fouling. In this type of application, it is usually better to employ two large diameter cyclones in series.

1.5.4 COARSE PARTICLES. When cyclones handle coarse particles, they are usually designed for low inlet velocities 5-10 feet per second (ft/sec). This is done to minimize erosion on the cyclone walls and to minimize breakdown of coarser particles that would normally be separated, into particles too fine for collection.

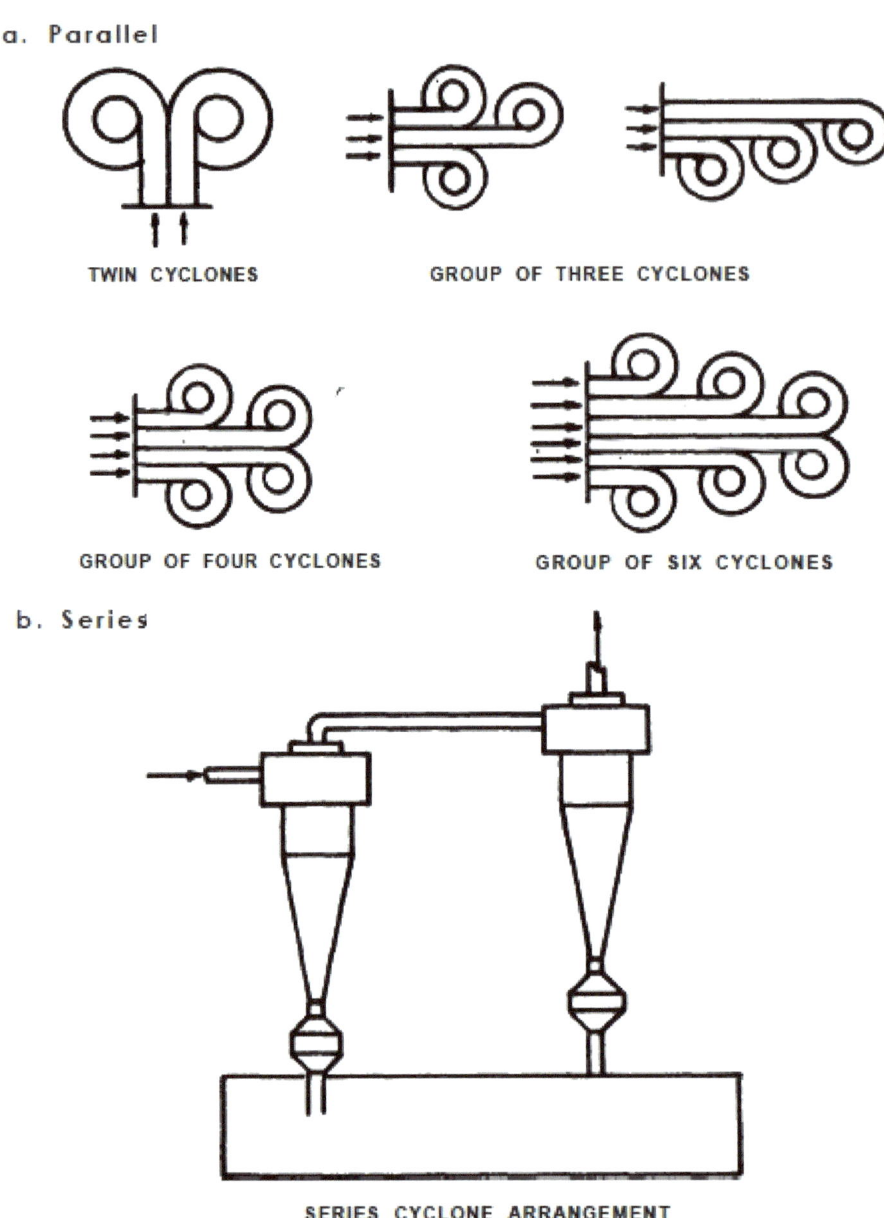

Figure 6-5

Parallel and series arrangements for cyclones.

© J. Paul Guyer   2021

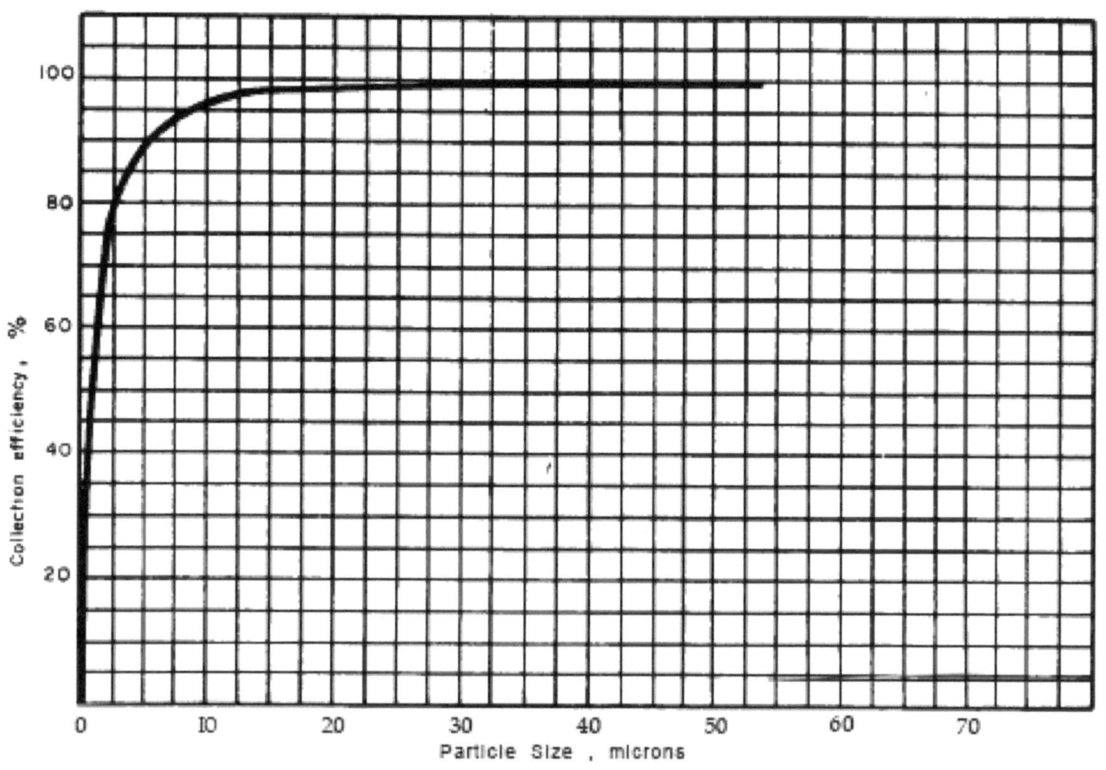

HIGH EFFICIENCY MULTIPLE CYCLONE UNITS (MULTICLONES)

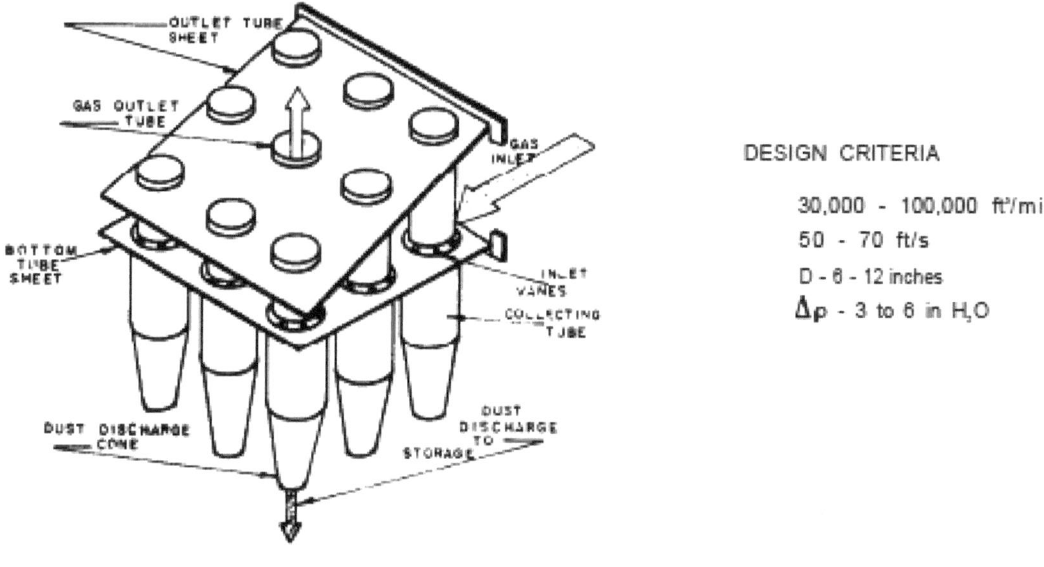

Figure 6-6

Efficiency curve and illustration of a multicyclone unit

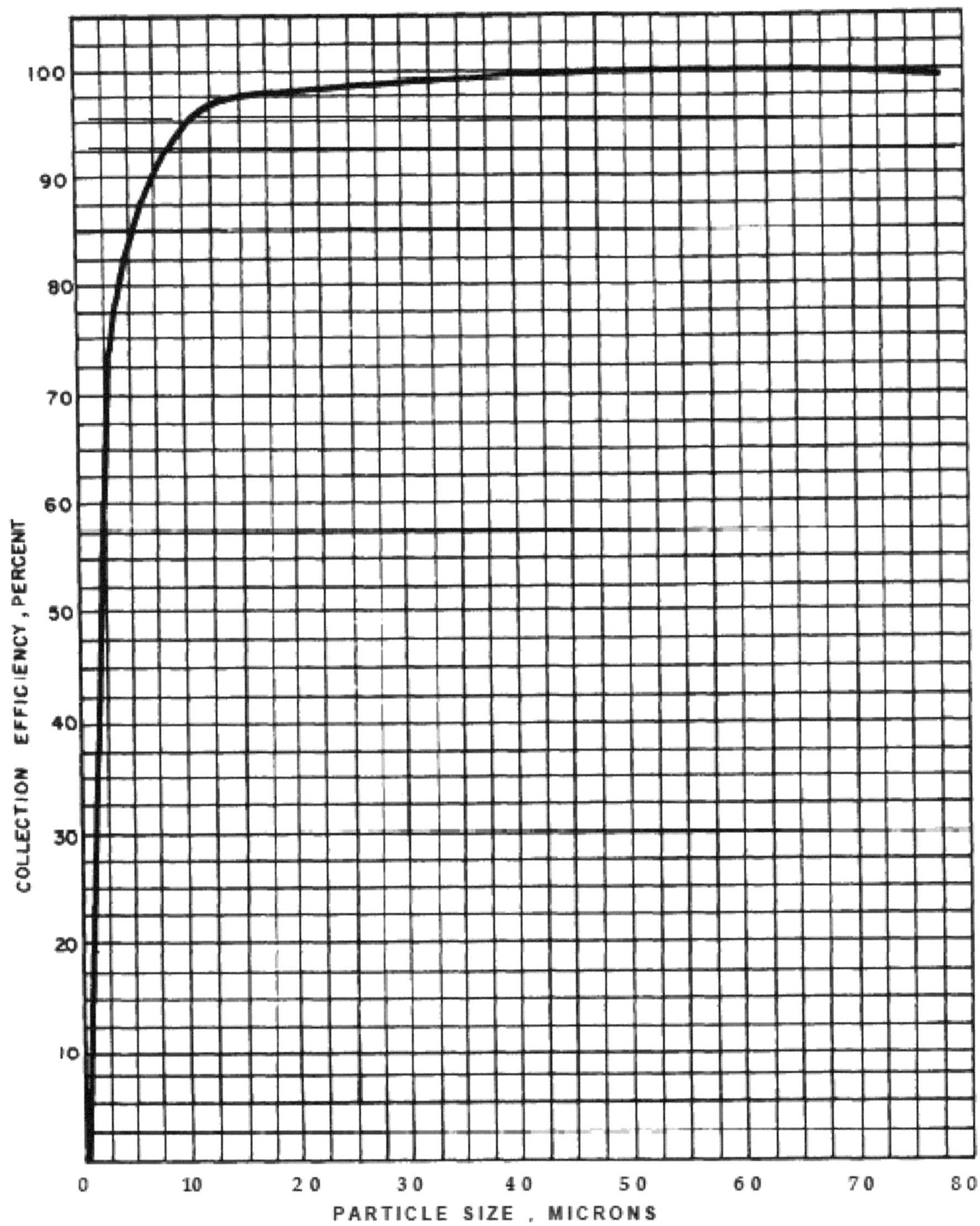

Figure 6-7

Size-efficiency curve for high-efficiency (long cone) irrigated cyclone.

© J. Paul Guyer   2021

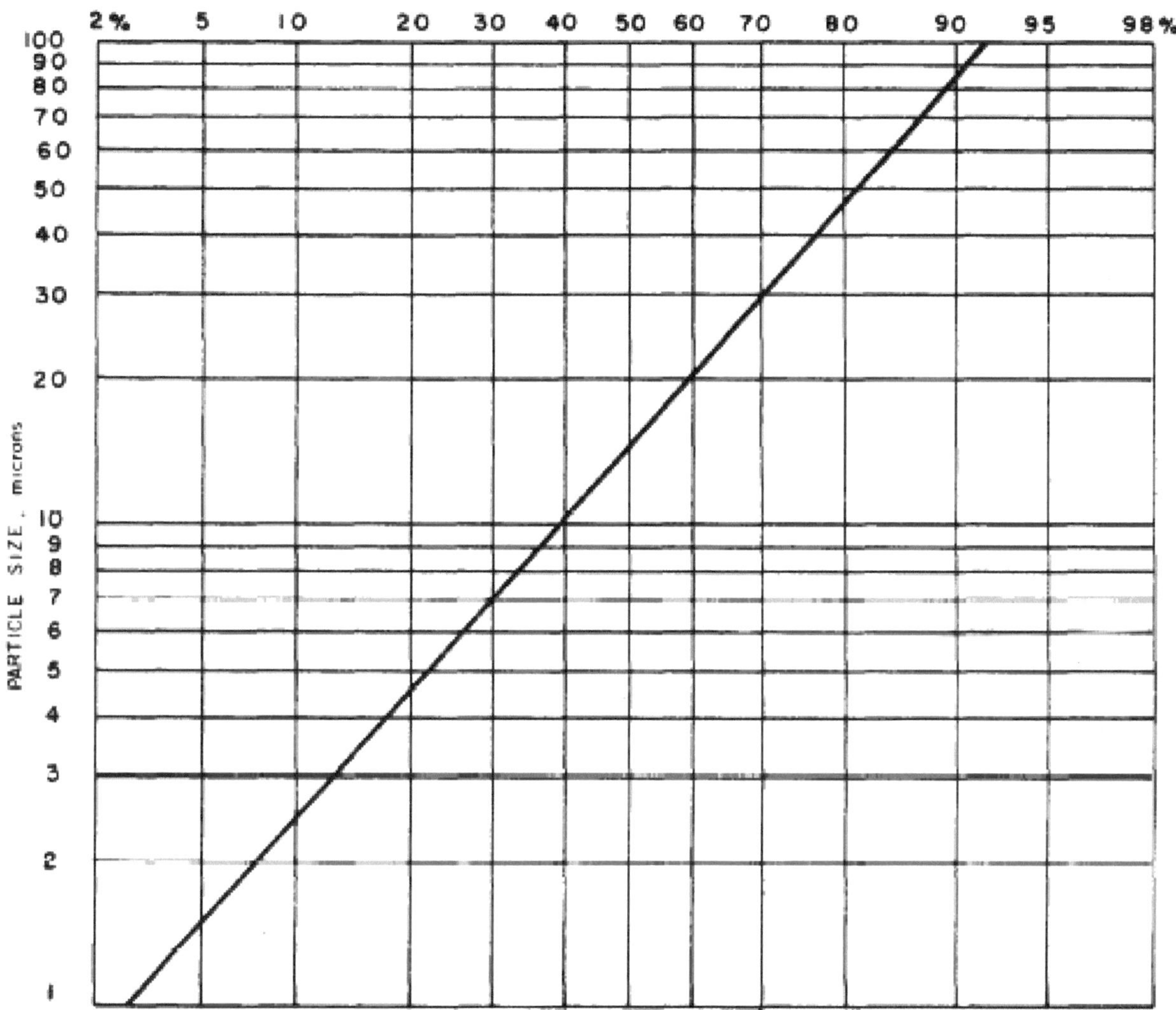

Figure 6-8

Particle size distribution curve (by weight) of particulate emitted from uncontrolled power

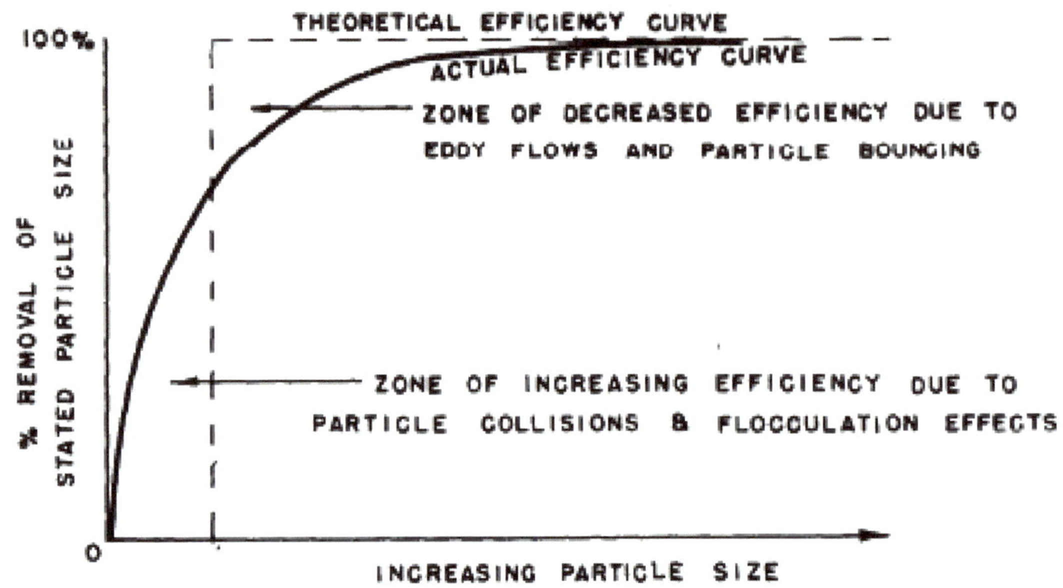

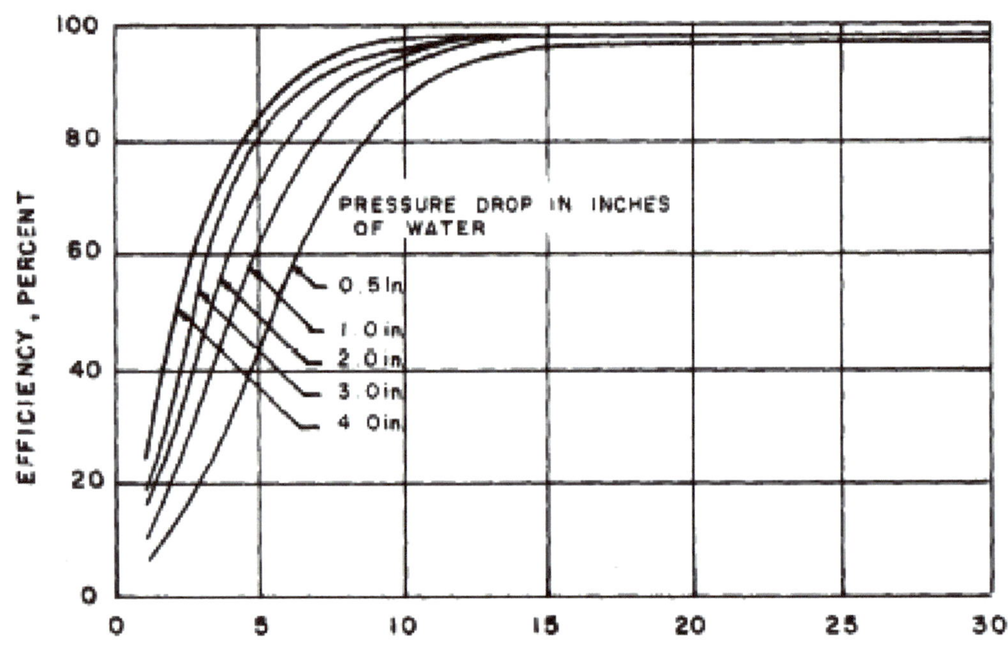

Figure 6-9

Fractional efficiency curves

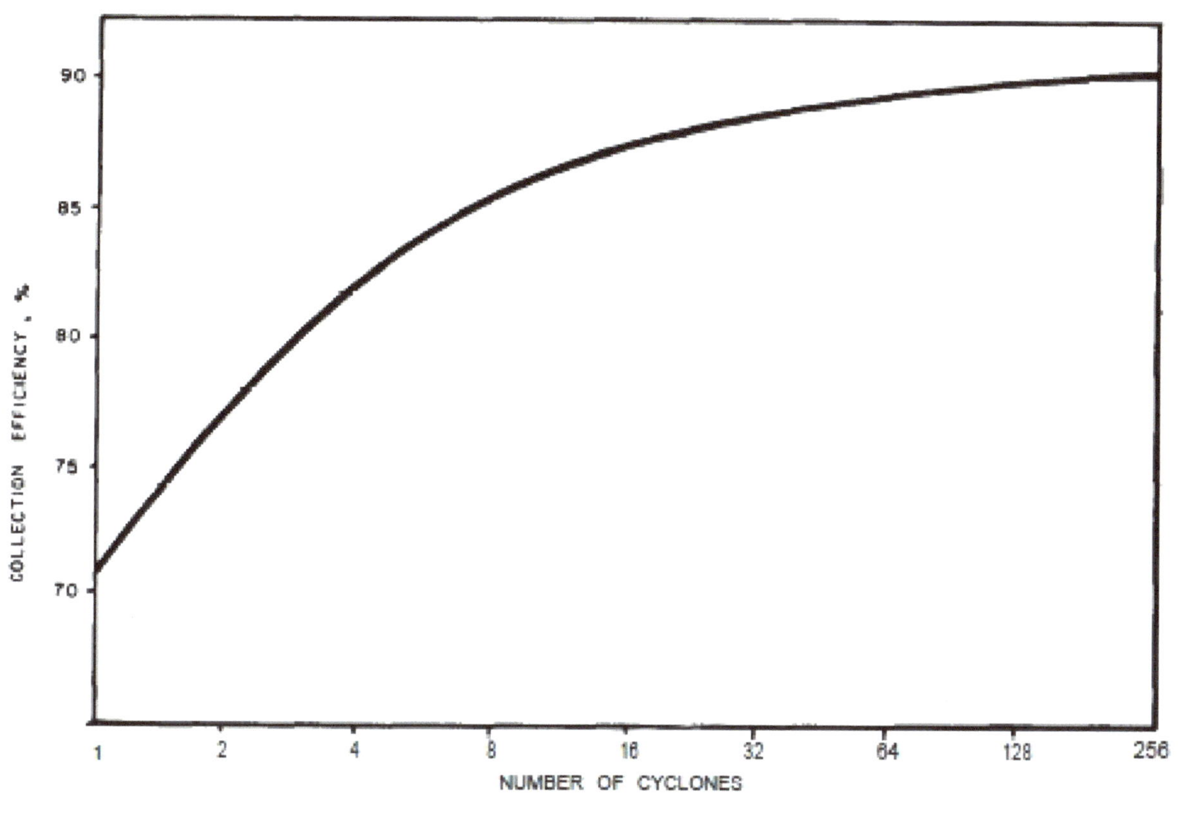

Figure 6-10

Effect of smaller cyclones in parallel on theoretical cyclone collection efficiency

1.5.5 LIMITED SPACE. In cases where cyclones must be erected in limited space, smaller diameter multicyclones have an obvious space advantage over larger diameter units. Small cyclones also have the advantage of increased efficiency over a single unit handling the same gas capacity, although this advantage is sometimes lost by uneven gas distribution to each unit with resultant fouling of some elements.

1.6 CYCLONE PERFORMANCE

1.6.1 COLLECTION EFFICIENCY AND PRESSURE DROP. For any given cyclone it is desirable to have as high a collection efficiency and as low a pressure drop as possible. Unfortunately, changes in design or operating variables which tend to increase collection efficiency also tend to increase pressure drop at a greater rate than the collection efficiency. Efficiency will increase with an increase in particle size, particle density, gas inlet velocity, cyclone body or cone length, and the ratio of body diameter to gas outlet diameter. Decreased efficiency is caused by an increase in gas viscosity, gas density, cyclone diameter, gas outlet diameter, and inlet widths or area. The effect on theoretical collection efficiency of changing the dimensions of an 8 inch diameter cyclones is shown in figure 6-11. The effects of changing gas inlet velocity, grain loading, particle specific gravity, gas viscosity, and particle size distribution on a 50 inch diameter cyclone are shown in figures 6-12 and 6-13. These figures illustrate the dependence of cyclone collection efficiency on those variables and the importance of maintaining proper gas inlet conditions.

1.6.2 FIELD PERFORMANCE. The actual in-field performance of cyclone units will vary because of changes in operating conditions such as dust load and gas flow. Table 6-2 illustrates the optimum expected performance of cyclone units for particulate removal application in combustion processes.

Source	Cyclones	
	Conventional	High Efficiency
	percent removed	
Coal tied:		
1. Spreader, chain grate, and vibrating stokers	70-80	80-90
2. Other stokers	75-80	90-95
3. Cyclone furnances	20-30	30-40
4. Other pulverized coal units	40-60	65-75

© J. Paul Guyer   2021

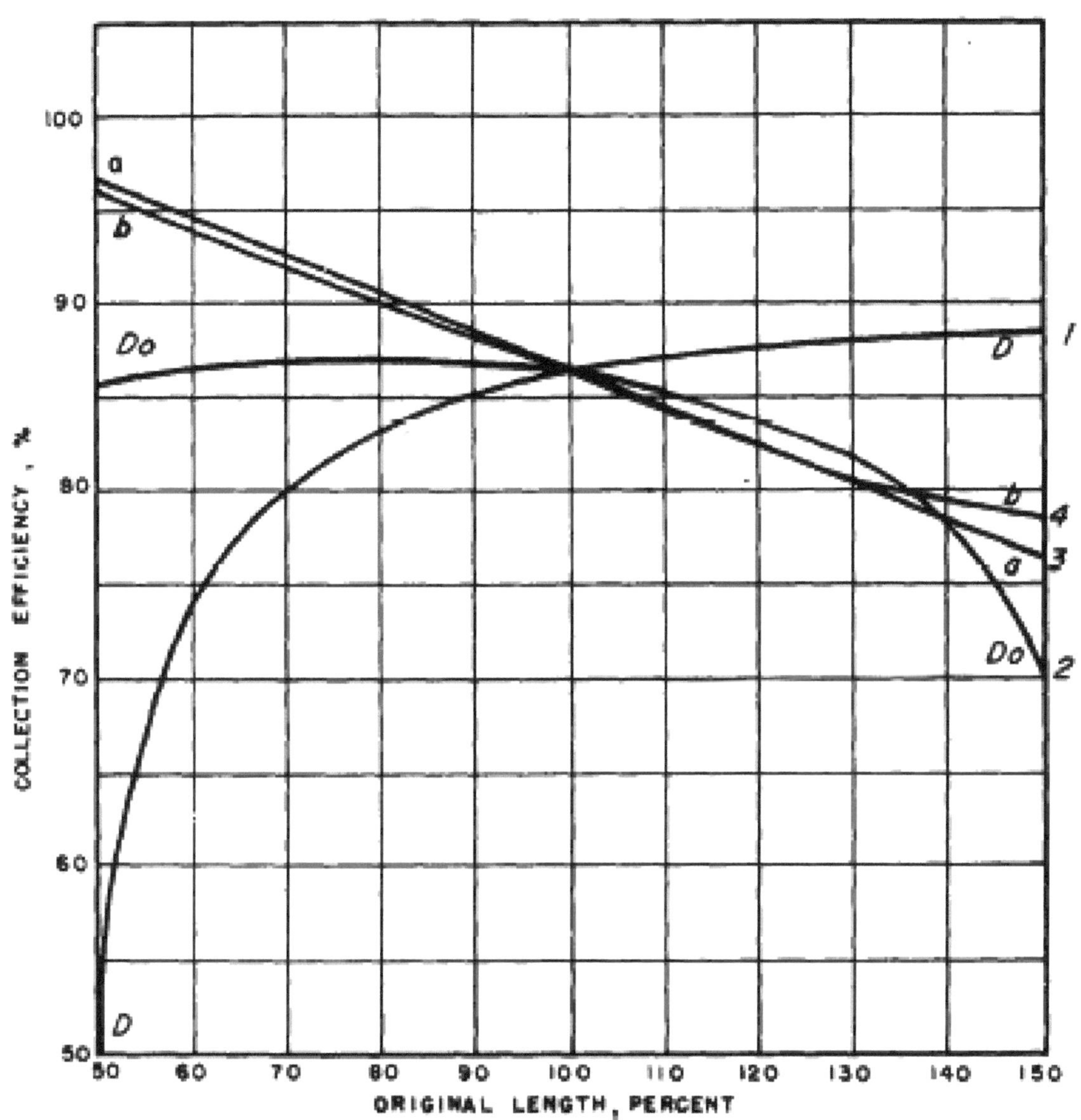

Curve 1 = cyclone body diameter
Curve 2 = cyclone gas outlet diameter
Curve 3 = cyclone inlet width
Curve 4 = cyclone inlet height

Figure 6-11

Effect of altering individual cyclone dimensions on theoretical collection efficiency for an

8-inch diameter cyclone

1.7 CYCLONE OPERATION

1.7.1 EROSION. Erosion in cyclones is caused by impingement and rubbing of dust on the cyclone walls. Erosion becomes increasingly worse with high dust loading, high inlet velocities, larger particle size, and more abrasive dust particles. Any defect in cyclone design or operation which tends to concentrate dust moving at high velocity will accelerate erosion. The areas most subject to erosive wear are opposite the inlet, along lateral or longitudinal weld seams on the cyclone walls, near the cone bottom where gases reverse their axial flow, and at mis-matched flange seams on the inlet or dust outlet ducting. Surface irregularities at welded joints and the annealed softening of the adjacent metal at the weld will induce rapid wear. The use of welded seams should be kept to a minimum and heat treated to maintain metal hardness. Continuous and effective removal of dust in the dust outlet region must be maintained in order to eliminate a high circulating dust load and resultant erosion. The cyclone area most subject to erosion is opposite the gas inlet where large incoming dust particles are thrown against the wall, and in the lower areas of the cone. Erosion in this area may be minimized by use of abrasion resistant metal. Often provisions are made from removable linings which are mounted flush with the inside surface of the shell. Erosion resistant linings of troweled or cast refractory are also used. Dust particles below the 5 to 10 micron range do not cause appreciable erosion because they possess little mass and momentum. Erosion is accelerated at inlet velocities above approximately 75 ft/sec.

1.7.2 FOULING. Decreased collection or by buildup of materials on the cyclone wall. Dust outlets become plugged by large pieces of extraneous material in the system, by overfilling of the dust bin, or by the breakoff of materials caked on the cyclone walls. The buildup of sticky materials on the cyclone walls is primarily a function of the dust properties. The finer or softer the dust, the greater is the tendency to cake on the walls. Condensation of moisture on the walls will contribute to dust accumulations. The collector should therefore be insulated to keep the surface temperature above the flue gas dew point. Wall buildup can generally be minimized by keeping the gas inlet velocity above 50 ft/sec.

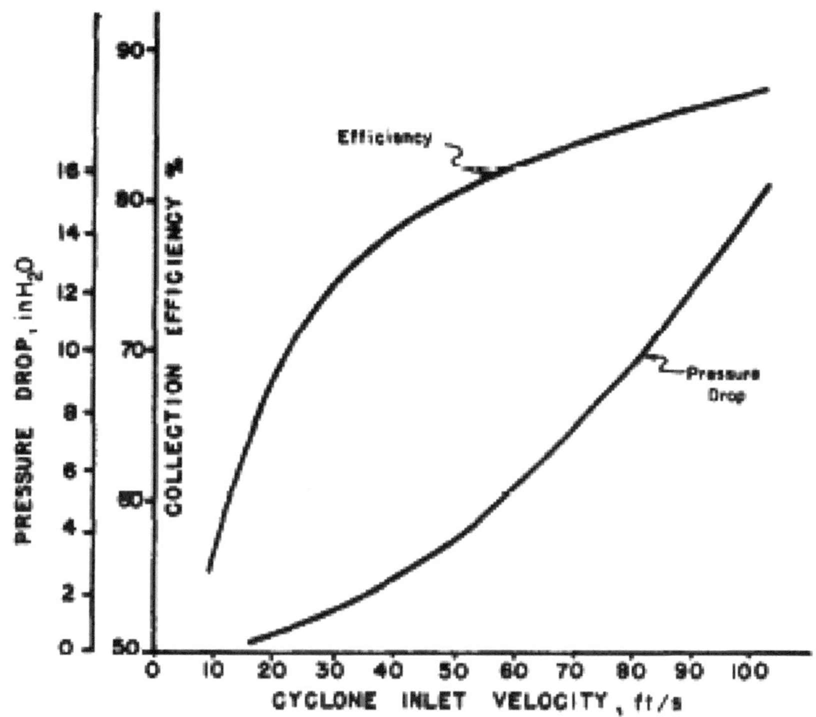

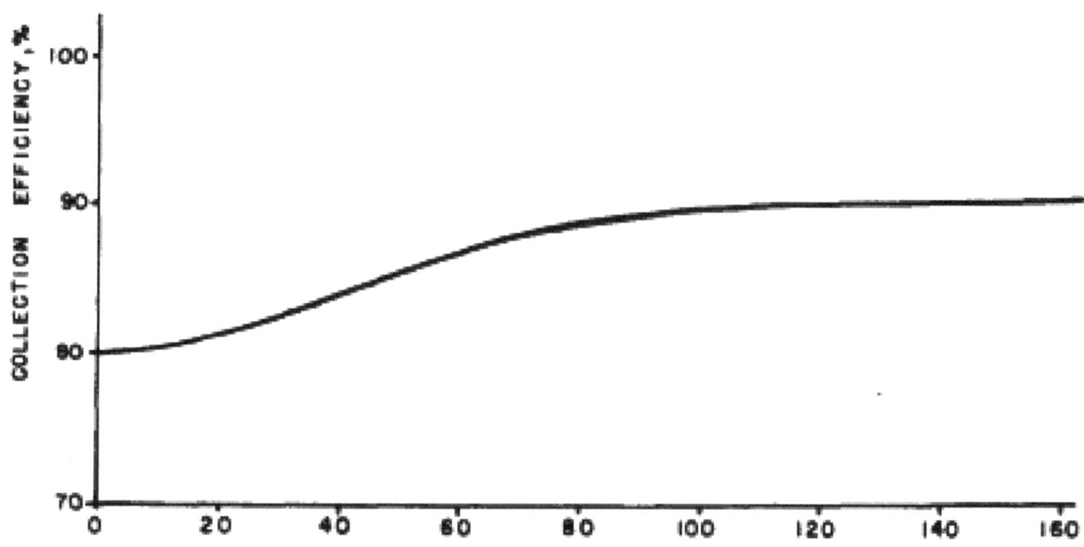

Figure 6-12

Effects of cyclone inlet velocity and grain loading

© J. Paul Guyer   2021   21

1.7.3 CORROSION. Cyclones handling gases containing sulfur oxides or hydrogen chloride are subject to acid corrosion. Acids will form when operating at low gas temperatures, or when the dust hopper may be cool enough to allow condensation of moisture. Corrosion is usually first observed in the hopper or between bolted sections of the cyclone inlet or outlet plenum spaces where gasketing material is used and cool ambient air can infiltrate. Corrosion at joints can be minimized by using welded sections instead of bolted sections. Ductwork and hoppers should be insulated and in cold climates the hoppers should be in a weather protected enclosure. Heat tracing of the hoppers may be necessary.

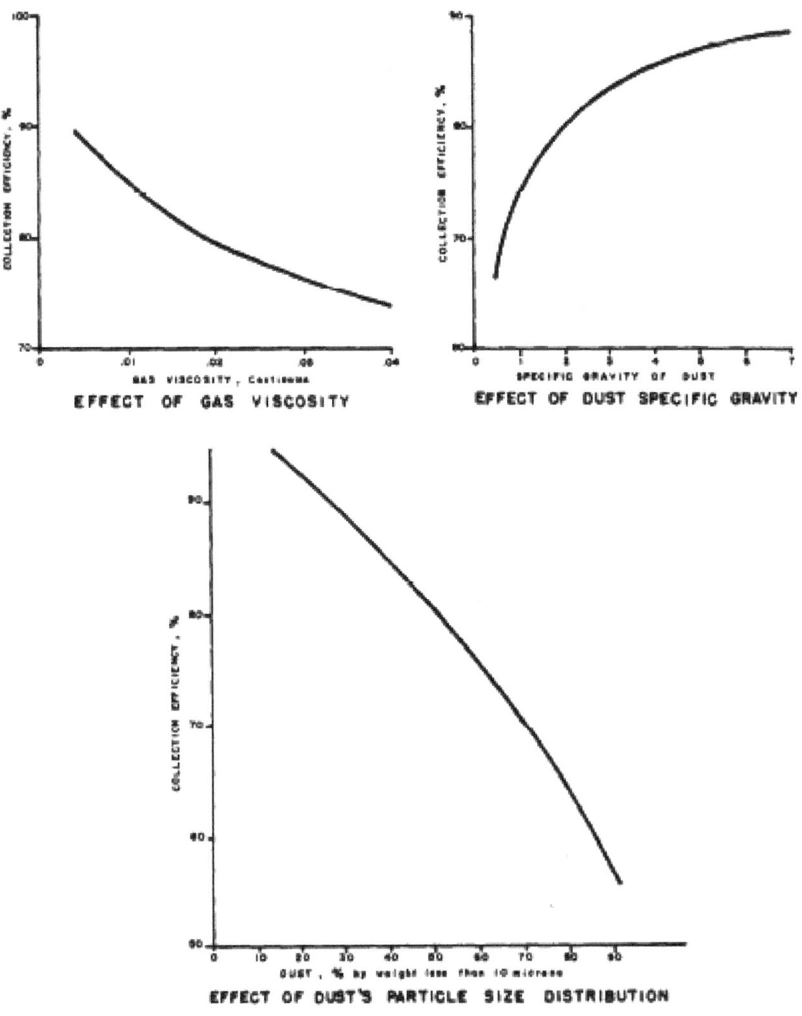

Figure 6-13.

Effects of gas viscosity, dust specific gravity, and dust particle size distribution.

1.7.4 DUST HOPPER DESIGN. A properly designed dust hopper should be air tight and large enough to prevent the dust level from reaching the cyclone dust outlets. Dust hoppers are usually conical or pyramidal in shape and are designed to prevent dust buildup against the walls. All designs should include a means of continuous removal of dust from the hopper to a storage bin, with an adequate alarm system to indicate a malfunction. Bin level alarms are frequently used for this purpose. On negative pressure systems, hoppers and removal system must be air tight. If hot unburned combustibles or char are present in the collected particulate, introduction of fresh air can cause a hopper fire. Pneumatic ash transport systems are not recommended for ash containing unburned combustibles or char for the same reason.

1.8 SELECTION OF MATERIALS

1.8.1 CONDITIONS. Cyclones can be constructed of a variety of types of metals. The type of materials specified is dependent upon the erosion characteristics of the dust, the corrosion characteristics of the gases, and the operating temperature of the cyclone. Generally, cyclones are constructed of mild steel or cast iron. (See para 7-5 for additional information on materials selection for pollution control systems).

1.8.2 EROSION. Erosion is the single most important criterion in specifying the materials for cyclone construction. Erosion life of a cyclone may be extended by using harder and thicker grades of steel. A stainless steel of 400 Brinell rating or better is normally chosen for cyclones subject to erosive conditions. When erosion is extreme, it is necessary to provide for replaceable liners in cyclone construction. Liners are made of hard stainless steels or erosion resistant refractory. In low temperature fly ash applications, cyclones of mild steel or iron can be used because dust loadings are generally too small to cause appreciable erosion. Cast iron is most often used in multicyclones in boiler service.

1.8.3 TEMPERATURE. Cyclones operated above 800 degrees Fahrenheit cannot be constructed of mild steels because the metal will creep and form ridges or buckled sections. Above 800 degrees Fahrenheit, nickel-copper bearing steel such as Monel is used to provide added strength. When temperatures are in excess of 1000 degrees Fahrenheit, nickel-chromium steel of the 400 series is used in conjunction with refractory linings. Silica carbide refractories provide excellent protection against erosion and high temperature deformation of the cyclone metal parts.

© J. Paul Guyer   2021

1.9 ADVANTAGES AND DISADVANTAGES

1.9.1 ADVANTAGES. The advantages of selecting cyclones over other particulate collection devices are:

- -No moving parts,
- -Easy to install and replace defective parts,
- -Constructed of a wide variety of materials,
- -Minimum space requirements,
- -Designed to handle severe service conditions of temperature, pressure, dust loading, erosion, corrosion, and plugging,
- -Can be designed to remove liquids from gas,
- -Low capital costs,
- -Low maintenance costs.

1.9.2 DISADVANTAGES. The disadvantages of selecting cyclones over other particulate collection devices are:

- -Lower collection efficiency,
- -Higher collection efficiencies (90-95 percent) only at high pressure drops (6 inches, water gauge),
- -Collection efficiency sensitive to changes in gas flow, dust load, and particle size distribution,
- -Medium to high operating costs.

CHAPTER 2
FABRIC FILTERS

2.1 FABRIC FILTRATION. Fabric filters are used to remove particles from a gas
stream. Fabric filters are made of a woven or felted material in the shape of a cylindrical
bag or a flat supported envelope. These elements are contained in a housing which has
gas inlet and outlet connections, a dust collection hopper, and a cleaning mechanism
for periodic removal of the collected dust from the fabric. In operation, dust laden gas
flows through the filters, which remove the particles from the gas stream. A typical fabric
filter system (baghouse) is illustrated in figure 9-1.

2.2 TYPES OF FILTERING SYSTEMS. The mechanisms of fabric filtration are identical regardless of variations in equipment structure and design. In all cases, particulates are filtered from the gas stream as the gas passes through a deposited dust matrix, supported on a fabric media. The dust is removed from the fabric periodically by one of the available cleaning methods. This basic process may be carried out by many different types of fabric filters with a variety of equipment designs. Filtering systems are differentiated by housing design, filter arrangement, and filter cleaning method.

2.2.1 HOUSING DESIGN. There are two basic housing configurations which apply to boiler and incinerator flue gas cleaning. These are closed pressure, and closed suction.

2.2.1.1 THE CLOSED PRESSURE BAGHOUSE is a completely closed unit having the fan located on the dirty side of the system. Toxic gases and gases with high dew points are handled in this type of baghouse. Fan maintenance problems arise due to the fact that the fan is in the dirty gas stream before the baghouse. The floor of the unit is closed and the hoppers are insulated. A closed pressure baghouse is illustrated in figure 9-2.

2.2.1.2 THE CLOSED SUCTION is the most expensive type of baghouse, with the fan being located on the clean gas side. The closed suction baghouse is an all-welded, air-tight structure. The floor is closed, and the walls and hopper are insulated. Fan maintenance is less than with the pressure type, but inspection of bags is more difficult. A closed suction system is illustrated in figure 9-2.

2.2.2 FILTER SHAPE AND ARRANGEMENTS.

2.2.2.1 THE CYLINDRICAL FILTER IS the most common filter shape used in fabric filtration. The principal advantage of a cylindrical filter is that it can be made very long. This maximizes total cloth area per square foot of floor space. Cylindrical filters are arranged to accommodate each of the basic flow configurations shown in figure 9-3.

2.2.2.2 A PANEL TYPE FILTER consists of flat areas of cloth stretched over an adjustable frame. (See figure 9-3.) Flow directions are usually horizontal. Panel filters allow 20 to 40 percent more cloth per cubic foot of collector volume and panels may be brushed down if dust build-up occurs. However, panel-type filters are not widely used in boiler and incinerator applications.

2.2.2.3 CLEANING METHODS. A fabric cleaning mechanism must impart enough energy to the cloth to overcome particle adhering forces without damaging the cloth, disturbing particle deposits in the hopper, or removing too much of the residual dust deposit on the filter. The cleaning period should be much shorter than the filtering period. The correct choice of cleaning method for a particular application will greatly enhance the performance of the fabric filter system. An incorrectly matched cleaning method can result in high pressure drops, low collection efficiency, or decreased bag life. A performance comparison of the various cleaning methods is given in table 9-I.

2.2.2.3.1 MECHANICAL SHAKE. Some baghouses employ a type of mechanical shaking mechanism for cleaning. Bags are usually shaken from the upper fastenings, producing vertical, horizontal, or a combination of motions, on the bag. All bags in a compartment may be fastened to a common framework, or rows of bags are attached to a common rocking shaft. After the bags have been shaken, loosened dust is allowed to settle before filtration is resumed. The entire cleaning cycle may take from 30 seconds to a few minutes. Some designs incorporate a slight reversal of gas flow to aid in dust cake removal and settling, as any slight flow in the direction of normal filtration will greatly reduce the effectiveness of cleaning. For this reason a positive sealing type valve is recommended for baghouse inlet and outlet. Shaker baghouses are normally used in small capacity systems or systems with a large number of filtering compartments.

© J. Paul Guyer   2021

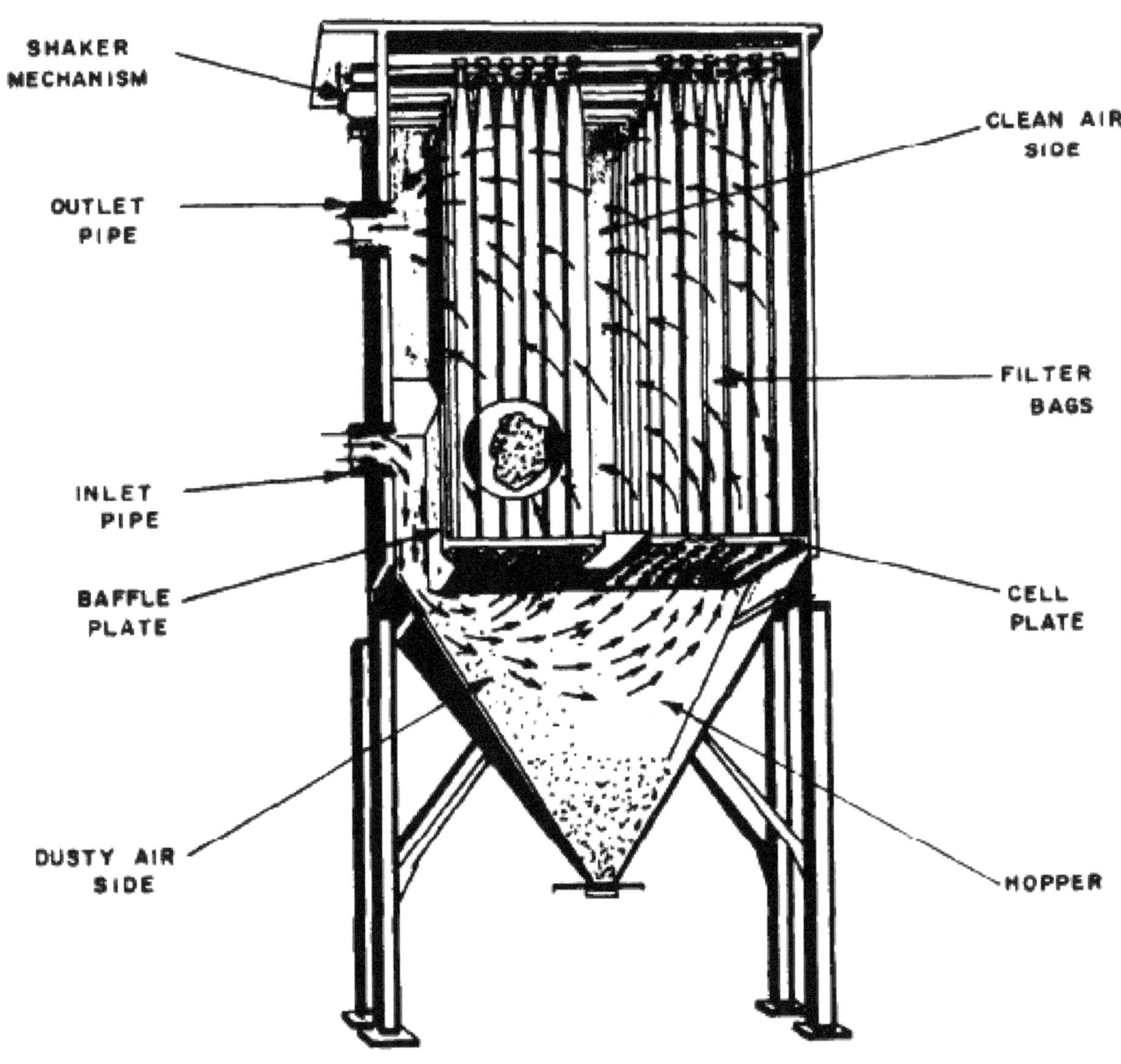

Figure 9-1

Typical baghouse (shake clean)

2.2.2.3.2 REVERSE FLOW WITHOUT BAG COLLAPSE. This cleaning method is used with a dust that releases fairly easily from the fabric. (See figure 9-4). A low pressure reversal of flow is all that is necessary to remove deposited dust from fabric. To minimize flexure and wear, the fabric is supported by a metal grid, mesh, or rings, sewn into the bag. Any flow that is reversed through the filter must be prefiltered. This results in increased total flow, requiring a greater cloth area, and producing a higher filtering velocity. This net increase in flow is normally less than 10 percent. Reverse pressures range from 125 pounds/square inch (lb/in2) down to a few inches, water gauge. The gentle cleaning action of reverse flow allows the use of glass fabric bags in high-temperature applications.

2.2.2.3.3 REVERSE FLOW WITH BAG COLLAPSE. Even though flexure can be detrimental to the bag, it is frequently utilized in order to increase the effectiveness of cleaning in a reverse baghouse. Filter bags collecting dust on the inside of the fabric are collapsed by a burst of reverse air which snaps the dust cake from the cloth surface. The bags do not collapse completely but form a cloverleaf type pattern. Collapse cleaning uses the same equipment arrangement as reverse flow without bag collapse. One design sends a short pulse of air down the inside of the bag, along with the reverse flow, to produce increased flexure and cleaning as is illustrated in figure 9-5. The principal disadvantage of flexural cleaning is the increased fabric wear. If the dust cake fails to be removed completely, the bag will stiffen in that area and cause wear in adjacent areas during cleaning.

2.2.2.3.4 REVERSE-FLOW HEATING. With a reverse flow cleaning system it may be necessary to have a reverse flow heating system. This system would be used to keep the gas temperatures in the baghouse above the acid dew point during the cleaning cycle.

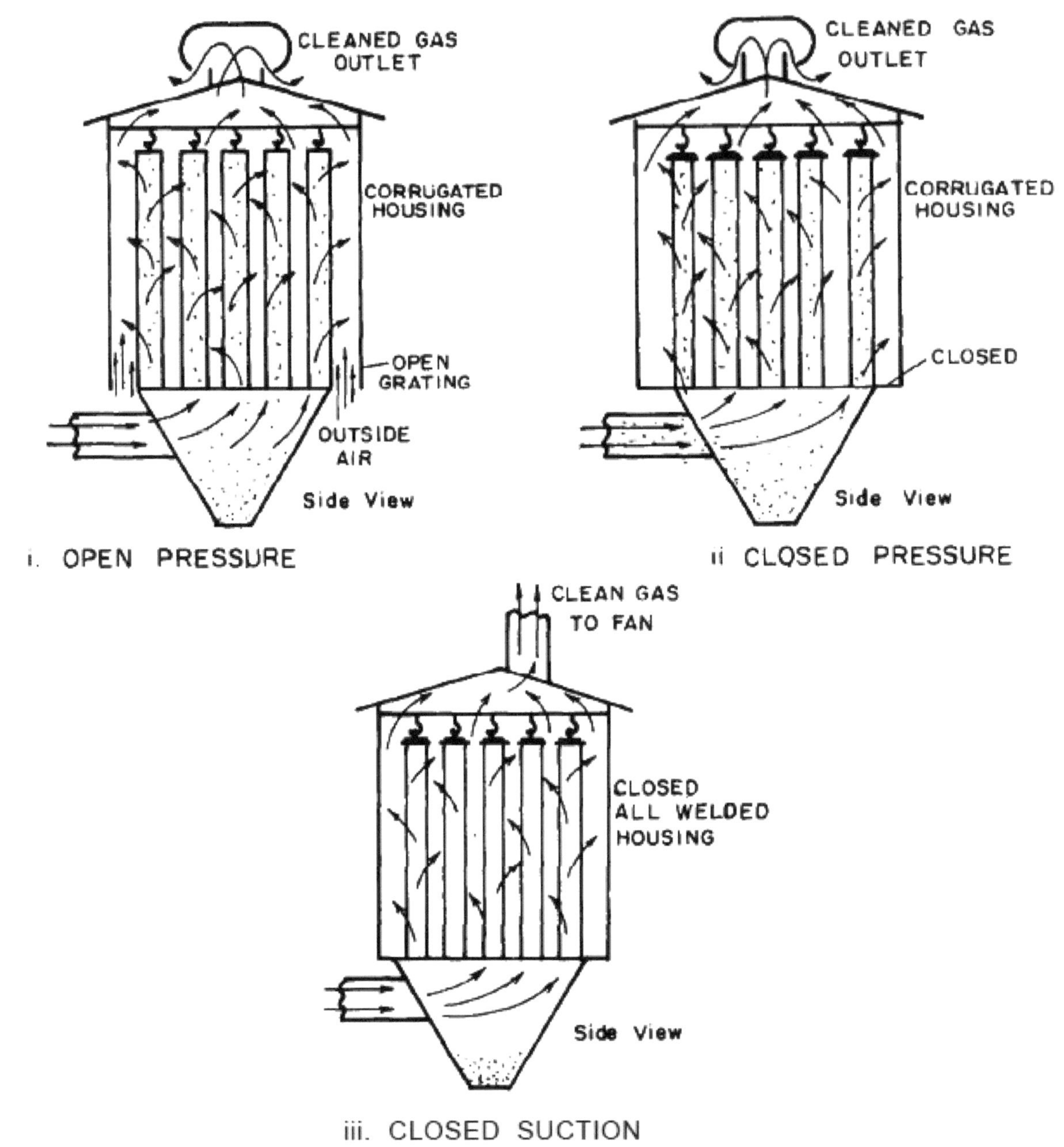

Figure 9-2

Fabric filter housing design

2.2.2.3.5 PULSE-JET. A pulse jet system is illustrated in figure 9-6. A short blast of air at 29 to 100 lb/ in, is directed into the top of the filter. This blast is usually sent through a venturi which increases the shock effect. As the pulse starts down the filter tube, more air is drawn in through the top. This combination causes the flow within the bag to temporarily reverse, bulges the fabric, and releases the dust cake from the outside of the filter tube. The whole process occurs in a fraction of a second which enables a virtually continuous filtering flow. Filter elements can be pulsed individually, or in rows. With a multi-compartment baghouse, a whole section may be pulsed at one time through a single venturi. The pulse produces less fabric motion than in shaking and also allows tighter bag spacing. A pulse-jet cleaning system requires no moving parts for cleaning and is designed to handle high gas flows per square foot of cloth area (air to cloth ratio). However, this system requires a compressed air system with a timer mechanism and control air solenoid valve for automatic cyclic cleaning. Pulse-jet baghouses are used when dust concentrations are high and continuous filtering is needed.

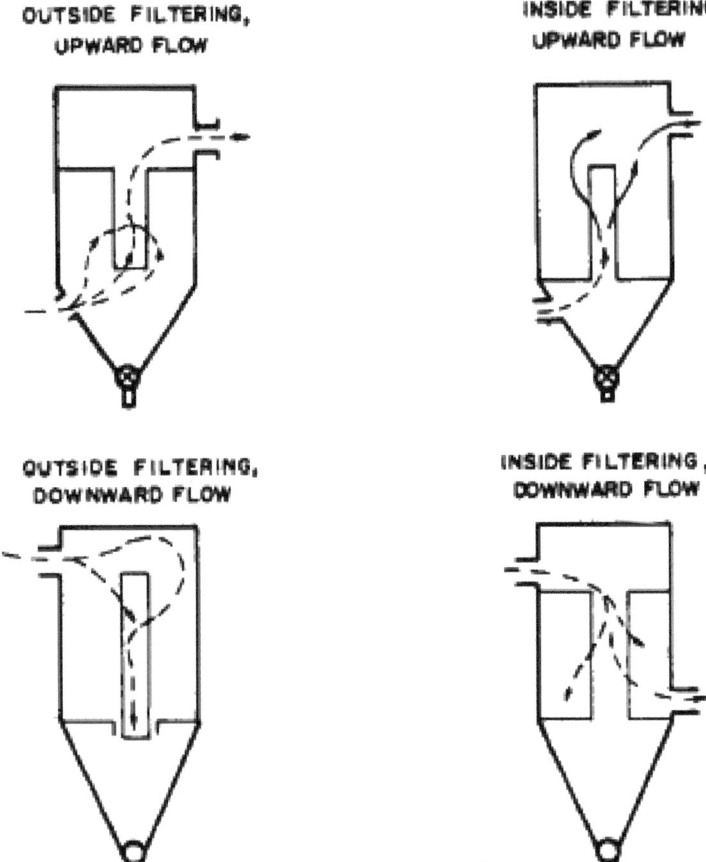

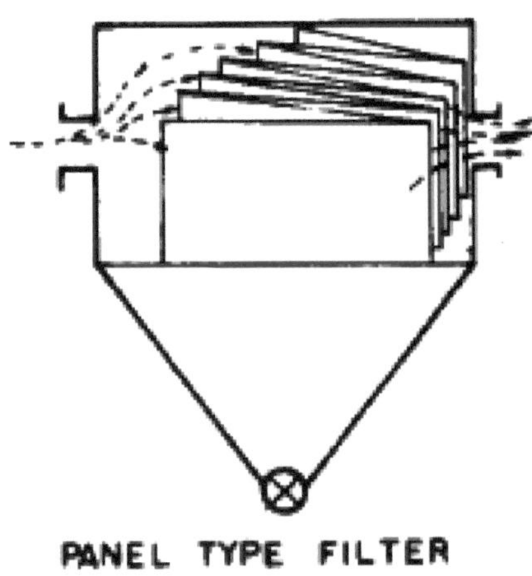

Figure 9-3

Filter shape and arrangement

2.3 FABRIC CHARACTERISTICS AND SELECTION. Fabric filter performance depends greatly upon the correct selection of a fabric. A fabric must be able to efficiently collect a specific dust, be compatible with the gas medium flowing through it, and be able to release the dust easily when cleaned. Fiber, yarn structure, and other fabric parameters will affect fabric performance. At the present time, the prediction of fabric pressure drop, collection efficiency, and fabric life is determined from past performance. It is generally accepted practice to rely on the experience of the manufacturer in selecting a fabric for a specific condition. However, the important fabric parameters are defined below to aid the user in understanding the significance of the fabric media in filtration.

2.3.1 FABRIC TYPE. The two basic types of fabric used in filtration are woven and felted. The woven fabric acts as a support on which a layer of dust is collected which forms a microporous layer and removes particles from the gas stream efficiently. A felted material consists of a matrix of closely spaced fibers which collect particles within its structure, and also utilizes the filter cake for further sieving. Filtering velocities for woven fabrics are generally lower than felts because of the necessity of rebuilding the cake media after each cleaning cycle. It is necessary that woven fabrics not be over-cleaned, as this will eliminate the residual dust accumulation that insures rapid formation of the filter cake and high collection efficiencies. Felts operate with less filter cake. This necessitates more frequent cleaning with a higher cleaning energy applied. Woven products, usually more flexible than felts, may be shaken or flexed for cleaning. Felts are usually back-washed with higher pressure differential air and are mainly used in pulsejet baghouses. However, felted bags do not function well in the collection of fines because the very fine particles become embedded in the felt and are difficult to remove in the cleaning cycle.

2.3.2 FIBER. The basic structural unit of cloth is the single fiber. Fiber must be selected to operate satisfactorily in the temperature and chemical environment of the gas being cleaned. Fiber strength and abrasion resistance are also necessary for extended filter life. The first materials used in fabric collectors were natural fibers such as cotton and wool. Those fibers have limited maximum operating temperatures (approximately 200 degrees Fahrenheit) and are susceptible to degradation from abrasion and acid condensation. Although natural fibers are still used for many applications, synthetic fibers such as acrylics, nylons, and Teflon have been increasingly applied because of their superior resistance to high temperatures and chemical attack (table 9-2).

2.3.2.1 ACRYLICS OFFER A GOOD COMBINATION of abrasion resistance and resistance to heat degradation under both wet and dry conditions. An outstanding characteristic of acrylics is the ability to withstand a hot acid environment, making them a good choice in the filtration of high sulfur-content exhaust gases.

2.3.2.2 AN OUTSTANDING NYLON FIBER available for fabric filters is Nomex, a proprietary fiber developed by DuPont for applications requiring good dimensional stability and heat resistance. Nomex nylon does not melt, but degrades rapidly in temperatures above 700 degrees Fahrenheit. Its effective operating limit is 450 degrees Fahrenheit. When in contact with steam or with small amounts of water vapor at elevated temperatures, Nomex exhibits a progressive loss of strength. However, it withstands these conditions better than other nylons and many other fibers. Because of Nomex's high abrasion resistance, it is used in filtration of abrasive dusts or wet abrasive solids and its good elasticity makes it ideal for applications where continuous flexing takes place. All nylon fabrics provide good cake discharge for work with sticky dusts.

System Type	Pressure Loss (in. of water)	Efficiency	Typical Cloth Type	Filter (cfm/ft^2 Cloth area)	Recommended Application	Advantages	Disadvantages
Shaker	3-6	99+%	Woven	1-5	Dust with good filter cleaning properties, intermittent collection	Relatively low initial investment	Bag failure increases with intensity and frequency of cleaning. Not suitable for fragile fabric and sticky dusts.
Reverse Flow	3-6	99+%	Woven	1-5	Dust with good filter cleaning properties, high temperature collection (incinerator fly ash) with glass bags	Low cloth attrition	Requires extra cloth capacity, additional fan and dampers.
Pulse Jet	3-6	99+%	Felted	5-20	Efficient for coal and oil fly ash collection.	Continuous flow pattern, low cloth attrition, unlimited dust concentration capability, no moving parts, no need for ducting dampers.	Compressed air required, high equipment cost.

Table 9-1

Performance comparison of fabric filter cleaning methods

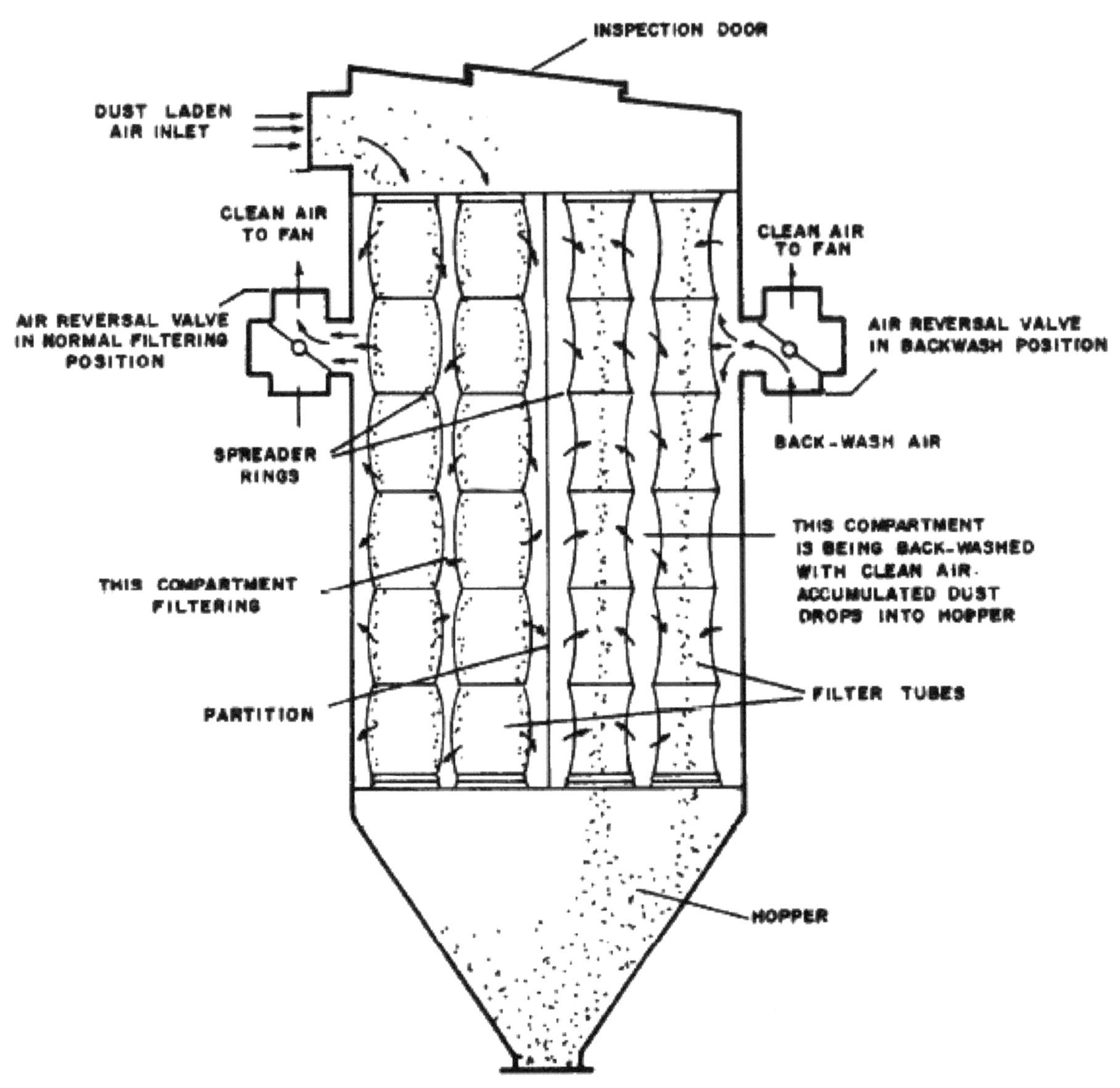

Figure 9-4

Reverse flow baghouse (without bag collapse)

2.3.2.3 TEFLON IS THE MOST chemically resistant fiber produced. The only substances known to react with this fiber are molten alkali metals, fluorine gas at high temperature and pressure, and carbon trifluoride. Teflon fibers have a very low coefficient of friction which produces excellent cake discharge properties. This fact, coupled with its chemical inertness and resistance to dry and moist heat degradation, make Teflon suitable for filtration and dust collection under severe conditions. Its major disadvantages are its poor abrasion resistance and high price. For these reasons, Teflon would be an economical choice only in an application where extreme conditions will shorten the service life of other filter fibers. It should be noted that the toxic gases produced by the decomposition of Teflon at high temperatures can pose a health hazard to personnel and they must be removed from the work area through ventilation.

2.3.3 YARN TYPE. Performance characteristics of filter cloth depend not only on fiber material, but also on the way the fibers are put together in forming the yarn. Yarns are generally classified as staple (spun) or filament.

2.3.3.1 FILAMENT YARNS SHOW BETTER release characteristics for certain dusts and fumes, especially with less vigorous cleaning methods.

2.3.3.2 STAPLE YARN GENERALLY PRODUCES a fabric of greater thickness and weight with high permeability to air flow. Certain fumes or dusts undergoing a change of state may condense on fiber ends and become harder to remove from the fabric.

2.3.4 WEAVE. The weave of a fabric is an important characteristic which affects filtration performance. The three basic weaves are plain, twill and satin.

2.3.4.1 PLAIN WEAVE is the simplest and least expensive method of fabric construction. It has a high thread count, is firm and wears well.

2.3.4.2 TWILL WEAVE gives the fabric greater porosity, greater pliability and resilience. For this reason twill weaves are commonly used where strong construction is essential.

2.3.4.3 SATIN FABRICS drape very well because the fabric weight is heavier than in other weaves. The yarns are compacted which produces fabric body and lower porosity, and they are often used in baghouses operating at ambient temperatures.

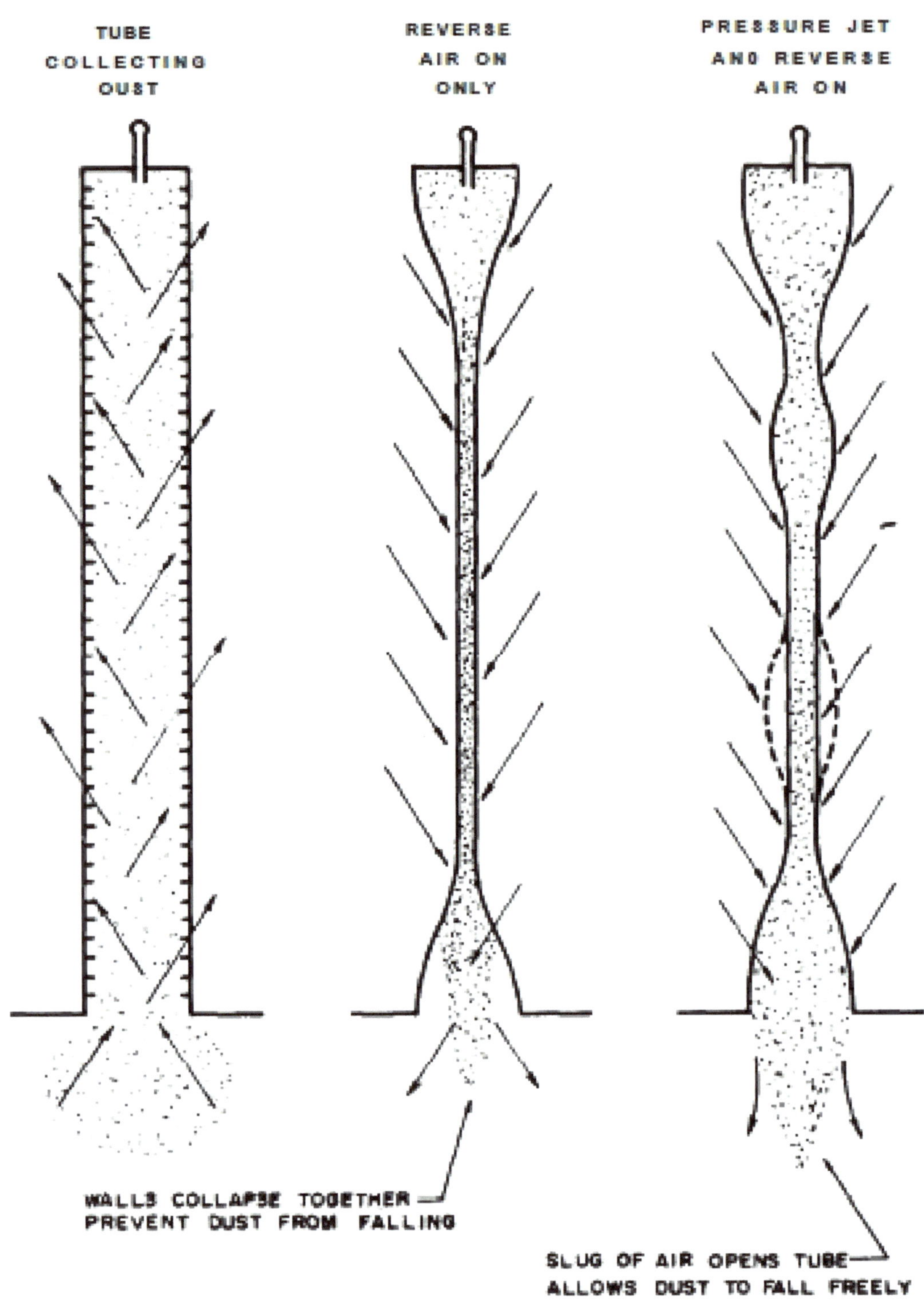

Figure 9-5.

Reverse flow cleaning (with bag collapse)

2.3.5 FINISH. Finishes are often applied to fabrics to lengthen fabric life. Cotton and wool can be treated to provide waterproofing, mothproofing, mildew-proofing and fireproofing. Synthetic fabrics can be heat-set to minimize internal stresses and enhance dimensional stability. Water repellents and antistatic agents may also be applied. Glass fabrics are lubricated with silicon or graphite to reduce the internal abrasion from brittle yarns. This has been found to greatly increase bag life in high temperature operations.

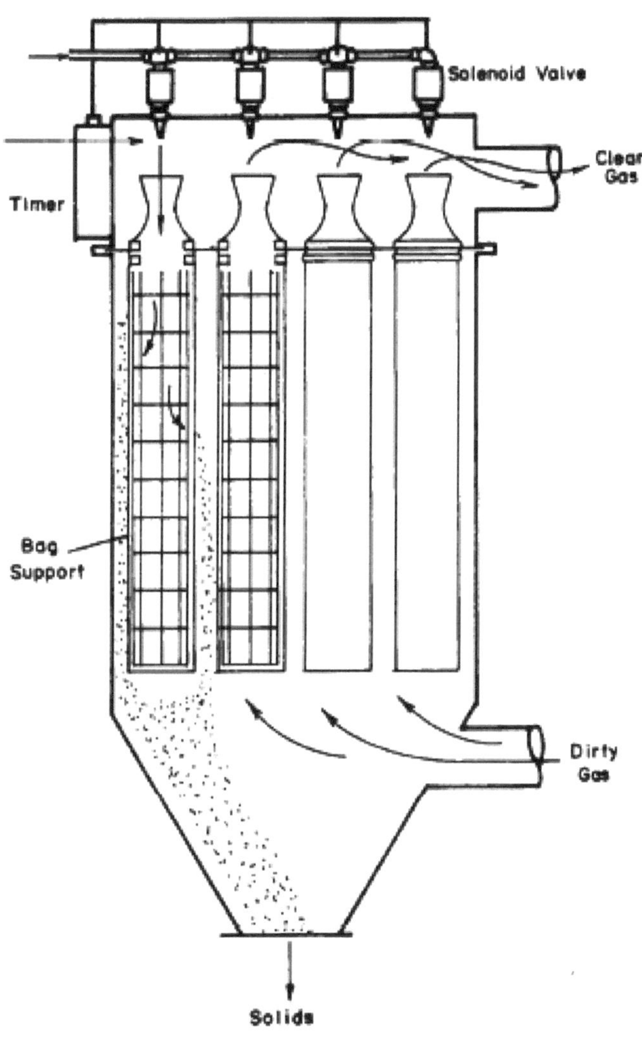

Figure 9-6

Pulse-jet baghouse

2.3.6 WEIGHT. Fabric weight is dependent upon the density of construction, and fiber or yarn weight. Heavier fabric construction yields lower permeability and increased strength.

2.4 MATERIALS AND CONSTRUCTION.

2.4.1 COLLECTOR HOUSING. Small unit collectors can be assembled at the factory or on location. Multi=compartment assemblies can be shipped by compartment or module (group of compartments), and assembled onsite. Field assembly is disadvantageous because of the need for insuring a good seal between panels, modules and flanges. Baghouse collector wall and ceiling panels are constructed of aluminum, corrugated steel, or concrete, the limitations being pressure, temperature, and corrosiveness of the effluent. The metal thickness must be adequate to withstand the pressure or vacuum within the baghouse and sufficient bracing should be provided. If insulation is needed, it can be placed between wall panels of adjacent compartments and applied to the outside of the structure. Pressure-relieving doors or panels should be included in the housing or ductwork to protect equipment if any explosive dust is being handled. An easy access to the baghouse interior must be provided for maintenance. Compartmented units have the advantage of being able to remain on-line while one section is out for maintenance. Walkways should be provided for access to all portions of the cleaning mechanism. Units with bags longer than 10 to 12 feet should be provided with walkways at the upper and lower bag attachment levels.

Generic Name Fiber Trade Name	Aramid Nomex	Glass Fiberglas	PTFE Teflon	Polyphenylene Sulfide Ryton	Polybenzi Midazole PBI	Metal Bekinox	Ceramic Nextel 312
Recommended continuous operation temperature (dry heat)	400°F	500°F	500°F	385°F	500°F	850°F	2,100°F
Water vapor saturated condition (moist heat)	350°F	500°F	500°F	375°F	500°F	750°F	2,199°F
Maximum (short time) operation temperature (dry heat)	450°F	550°F	550°F	450°F	650°F	950°F	2,600°F
Specific density	1.38	2.54	2.3	1.38	1.43	7.9	2.7
Relative moisture regain in % (at 68°F and 65% relative moisture	4.5	0	0	0.6	14	0	0
Supports combustion	No	No	No	No	No	No	No
Biological resistance (bacteria, mildew)	No Effect	No Effect	No Effect	No Effect	No Effect	No Effect	No Effect
Resistance to alkalies	Good	Fair	lent	Excellent	Good	Good	Good
Resistance to mineral acids	Fair	Very Good	Excel- lent	Excellent	Excellent	Very Good	Very Good
Resistance to organic acids	Fair +	Very Good	Excel- lent	Excellent	Excellent	Very Good	Very Good
Resistance to oxidizing agents	Poor	Excel- lent	Excel- lent	attacked by strong oxidiz- ing agents.	Fair	Very Good	Excellent
Resistance to organic solvents	Very Good	Very Good	Excel- lent	Excellent	Excellent	Very Good	Excellent

Table 9-2

Properties of fibers for high temperature dry filtration

2.4.2 HOPPER AND DISPOSAL EQUIPMENT. The dust-collection hopper of a baghouse can be constructed of the same material as the external housing. In small light duty, hoppers 16 gage metal is typical. However, metal wall thicknesses should be increased for larger baghouses and hopper dust weight. The walls of the hopper must be insulated and should have heaters if condensation might occur. The hopper sides should be sloped a minimum of 57 degrees to allow dust to flow freely. To prevent bridging of certain dusts, a greater hopper angle is needed, but continuous removal of the dust will also alleviate bridging. If dust bridging is a significant problem, vibrators or rappers may be installed on the outside of the hopper. The rapping mechanism can be electrically or pneumatically operated and the size of the hopper must be sufficient to hold the collected dust until it is removed. Overfilled hoppers may cause an increased dust load on the filter cloths and result in increased pressure drop across the collector assembly. Storage hoppers in baghouses which are under positive or negative pressure warrant the use of an air-lock valve for discharging dust. Since this will prevent re-entrainment of dust or dust blowout. A rotary air valve is best suited for this purpose.

2.4.3 FOR LOW SOLIDS FLOW, a manual device such as a slide gate, trip gate, or trickle valve may be used, however, sliding gates can only be operated when the compartment is shut down. For multi-compartmented units, screw conveyors, air slides, belt conveyors or bucket conveying systems are practical. When a screw conveyor or rotary valve is used, a rapper can be operated by a cam from the same motor.

2.5 AUXILIARY EQUIPMENT AND CONTROL SYSTEMS

2.5.1 INSTRUMENTATION. Optimum performance of a fabric filter system depends upon continuous control of gas temperature, system pressure drop, fabric pressure, gas volume, humidity, condensation, and dust levels in hoppers. Continuous measurements of fabric pressure drop, regardless of the collector size, should be provided. Pressure gages are usually provided by the filter manufacturer. With high and with variable dust loadings, correct fabric pressure drop is critical for proper operation and maintenance. Simple draft gages may be used for measuring fabric pressure drop, and they will also give the static pressures at various points within the system. Observation of key pressures within small systems, permits manual adjustment of gas flows and actuation of the cleaning mechanisms.

2.5.2 THE NUMBER AND DEGREE of sophistication of pressure- sensing devices is relative to the size and cost of the fabric filter system. High temperature filtration will require that the gas temperature not exceed the tolerance limits of the fabric and temperature displays are required to indicate whether necessary dilution air dampers or pre-cooling sprays are operating correctly. A well-instrumented fabric filter system protects the investment and decreases chances of malfunctions. It also enables the operating user to diagnose and correct minor problems without outside aid.

2.5.3 GAS PRECONDITIONING. Cooling the inlet gas to a fabric filter reduces the gas volume which then reduces required cloth area; extends fabric life by lowering the filtering temperature; and permits less expensive and durable materials to be used. Gas cooling is mandatory when the effluent temperature is greater than the maximum operating temperature of available fabrics. Three practical methods of gas cooling are radiation convection cooling, evaporation, and dilution.

2.5.2.1 RADIATION CONVECTION COOLING enables fluctuations in temperature, pressure, or flow to be dampened. Cooling is achieved by passing the gas through a duct or heat-transfer device and there is no increase in gas filtering volume. However, ducting costs, space requirements, and dust sedimentation are problems with this method.

2.5.2.2 EVAPORATIVE COOLING is achieved by injecting water into the gas stream ahead of the filtering system. This effectively reduces gas temperatures and allows close control of filtering temperatures. However, evaporation may account for partial dust removal and incomplete evaporation may cause wetting and chemical attack of the filter media. A visible stack plume may occur if gas temperatures are reduced near to or below the dew point. (3) Dilution cooling is achieved by mixing the gas steam with outside air. This method is inexpensive but increases filtered gas volume requiring an increase in baghouse size. It is possible the outside air which is added may also require conditioning to control dust and moisture content from ambient conditions.

2.6 ENERGY REQUIREMENTS. The primary energy requirement of baghouses is the power necessary to move gas through the filter. Resistance to gas flow arises from the pressure drop across the filter media and flow losses resulting from friction and turbulent effects. In small or moderately sized baghouses, energy required to drive the cleaning mechanism and dust disposal equipment is small, and may be considered negligible when compared with primary fan energy. If heating of reverse air is needed this will require additional energy.

2.7 APPLICATION.

2.7.1 INCINERATORS. Baghouses have not been widely used with incinerators for the following reasons:

2.7.1.1 MAXIMUM OPERATING TEMPERATURES for fabric filters have typically been in the range of 450 to 550 degrees Fahrenheit, which is below the flue gas temperature of most incinerator installations

2.7.1.2 COLLECTION OF CONDENSED TAR materials (typically emitted from incinerators) could lead to fabric plugging, high pressure drops, and loss of cleaning efficiency

2.7.1.3 PRESENCE OF CHLORINE AND MOISTURE in solid waste leads to the formation of hydrochloric acid in exhaust gases, which attacks fiberglass and most other filter media

2.7.1.4 METAL SUPPORTING FRAMES show distortion above 500 degrees Fahrenheit and chemical attack of the bags by iron and sulphur at temperatures greater than 400 degrees Fahrenheit contribute to early bag failure. Any fabric filtering systems designed for particulate control of incinerators should include:

- -fiberglass bags with silica, graphite, or teflon lubrication; or nylon and, teflon fabric bags for high temperature operation, or stainless steel fabric bags,
- -carefully controlled gas cooling to reduce high temperature fluctuations and keep the temperature above the acid dew point,
- -proper baghouse insulation and positive sealing against outside air infiltration. Reverse air should be heated to prevent condensation.

2.7.2 BOILERS. Electric utilities and industrial boilers primarily use electrostatic precipitators for air pollution control, but some installations have been shown to be successful with reverse air and pulse-jet baghouses. The primary problem encountered with baghouse applications is the presence of sulphur in the fuel which leads to the formation of acids from sulphur dioxide (SO2) and sulphur trioxide (SO3) in the exhaust gases. Injection of alkaline additives (such as dolomite and limestone) upstream of baghouse inlets can reduce SO2 present in the exhaust. Fabric filtering systems designed for particulate collection from boilers should: -operate at temperatures above the acid dew point,

- -employ a heated reverse air cleaning method,
- -be constructed of corrosion resistant material,
- -be insulated and employ internal heaters to prevent acid condensation when the installation is off-line.

2.7.3 SO$_2$ REMOVAL. The baghouse makes a good control device downstream of a spray dryer used for SO2 removal and can remove additional SO2 due to the passage of the flue-gas through unreacted lime collected on the bags.

2.7.4 WOOD REFUSE BOILER APPLICATIONS. It is not recommended that a baghouse be installed as a particulate collection device after a wood fired boiler. The possibility of a fire caused by the carryover of hot glowing particles is too great.

2.8 PERFORMANCE. Significant testing has shown that emissions from a fabric filter consist of particles less than 1 micron in diameter. Overall fabric filter collection efficiency is 99 percent or greater (on a weight basis). The optimum operating characteristics attainable with proper design of fabric filter systems are shown in table 9-3.

2.9 ADVANTAGES AND DISADVANTAGES

2.9.1 ADVANTAGES.

2.9.1.1 VERY HIGH COLLECTION efficiencies possible (99.9+ percent) with a wide range of inlet grain loadings and particle size variations. Within certain limits fabric collectors have a constancy of static pressure and efficiency, for a wider range of particle sizes and concentrations than any other type of single dust collector.

2.9.1.2 COLLECTION EFFICIENCY not affected by sulfur content of the combustion fuel as in ESPs.

2.9.1.3 REDUCED SENSITIVITY to particle size distribution.

2.9.1.4 NO HIGH VOLTAGE requirements.

2.9.1.5 FLAMMABLE dust may be collected.

2.9.1.6 USE OF SPECIAL FIBERS or filter aids enables submicron removal of smoke and fumes.

2.9.1.7 COLLECTORS AVAILABLE in a wide range of configurations, sizes, and inlet and outlet locations.

2.9.2 DISADVANTAGES.

2.9.2.1 FABRIC LIFE MAY BE substantially shortened in the presence of high acid or alkaline atmospheres, especially at elevated temperatures.

2.9.2.2 MAXIMUM OPERATING temperature is limited to 550 degrees Fahrenheit, unless special fabrics are used.

2.9.2.3 COLLECTION OF HYGROSCOPIC materials or condensation of moisture can lead to fabric plugging, loss of cleaning efficiency, large pressure losses.

2.9.2.4 CERTAIN DUSTS MAY require special fabric treatments to aid in reducing leakage or to assist in cake removal.

2.9.2.5 HIGH CONCENTRATIONS OF dust present an explosion hazard.

2.9.2.6 FABRIC BAGS TEND to burn or melt readily at temperature extremes.

Collection efficiency	99 percent+
Particle size collected	Greater than .5 microns
Pressure drop	.5-6 inches, water gauge
Maximum operating temperatures	550 degrees Fahrenheit peak and 500 degrees Fahrenheit continuous with common fabrics
Dust concentration handled in particulate collection	0-10 gr/ft^3
Gas volume	Unlimited
Cloth area	Several square feet to several thousand square feet
Filtering velocities	1-15 ft/min
Average bag life	18 months to 2 years

Table 9-3

Operating characteristics of fabric filters

CHAPTER 3
SCRUBBERS AND PRECIPITATORS

3.1 SCRUBBERS IN GENERAL. A scrubber utilizes a liquid to separate particulate or gaseous contaminants from gas. Separation is achieved through mass contact of the liquid and gas. Boiler emissions to be controlled include fly ash and sulfur oxides. Incinerator emissions to be controlled include fly ash, sulfur oxides and hydrogen chloride.

3.2 TYPES OF SCRUBBERS

3.2.1 LOW ENERGY SCRUBBERS. Low energy scrubbers are more efficient at gaseous removal than at particulate removal. A low energy scrubber utilizes a long liquid/gas contact time to promote mass transfer of gas. Low energy scrubbers depend on extended contact surface or interface between the gas and liquid streams to allow collection of particulate or gaseous emissions. (1) Plate-type scrubbers. A plate-type scrubber consists of a hollow vertical tower with one or more plates (trays) mounted transversely in the tower (figure 7-I). Gas comes in at the bottom of the tower, and must pass through perforations, valves, slots, or other openings in each plate before exiting from the top. Liquid is usually introduced at the top plate, and flows successively across each plate as it moves downward to the liquid exit at the bottom. Gas passing through the openings in each plate mixes with the liquid flowing over the plate. The gas and liquid contact allows the mass transfer or particle removal for which the plate scrubber was designed. Plate-type scrubbers have the ability to remove gaseous pollutants to any desired concentration provided a sufficient number of plates are used. They can also be used for particle collection with several sieve (perforated) plates combining to form a sieve-plate tower. In some designs, impingement baffles are placed a short distance above each perforation on a sieve plate, forming an impingement plate upon which particles are collected. The impingement baffles are below the level of liquid on the perforated plates and for this reason are continuously washed clean of collected particles. Particle collection efficiency is good for particles larger than one micron in diameter. Design pressure drop is about 1.5 inches of water for each plats.

3.2.2 PREFORMED SPRAY SCRUBBERS. A preformed spray scrubber (spray tower) is a device which collects particles or gases on liquid droplets and utilizes spray nozzles for liquid droplet atomization (figure 7-2). The sprays are directed into a chamber suitably shaped to conduct the gas through the atomized liquid droplets. Spray towers are designed for low pressure drop and high liquid consumption. They are the least expensive method for achieving gas absorption because of their simplicity of construction with few internals. The operating power cost is low because of the low gas pressure drop. Spray towers are most applicable to the removal of gases which have high liquid solubilities. Particle collection efficiency is good for particles larger than several microns in diameter. Pressure drops range from 1 to 6 inches, water gauge.

3.2.3 CENTRIFUGAL SCRUBBERS. Centrifugal scrubbers are cylindrical in shape, and impart a spinning motion to the gas passing through them. The spin may come from introducing gases to the scrubber tangentially or by directing the gas stream against stationary swirl vanes (figure 7-2). More often, sprays are directed through the rotating gas stream to catch particles by impaction upon the spray drops. Sprays can be directed outward from a central spray manifold or inward from the collector walls. Spray nozzles mounted on the wall are more easily serviced when made accessible from the outside of the scrubber. Centrifugal scrubbers are used for both gas absorption and particle collection and operate with a pressure drop ranging from 3 to 8 inches, water gauge. They are inefficient for the collection of particles less than one or two microns in diameter.

3.2.4 IMPINGEMENT AND ENTRAINMENT SCRUBBERS. Impingement and entrainment scrubbers employ a shell which holds liquid (figure 7-3). Gas introduced into a scrubber is directed over the surface of the liquid and atomizes some of the liquid into droplets. These droplets act as the particle collection and gas absorption surfaces. Impingement and entrainment scrubbers are most frequently used for particle collection of particles larger than several microns in diameter. Pressure drops range from 4 to 20 inches, water gauge. (5) Moving bed scrubbers. Moving bed scrubbers provide a zone of mobile packing consisting of plastic, glass, or marble spheres where gas and liquid can mix intimately (figure 7-3). A cylindrical shell holds a perforated plats on which the movable packing is placed. Gas passes upward through the perforated plate and/or down over the top of the moving bed. Gas velocities are sufficient to move the packing material when the scrubber is operating which aids in making the bed turbulent and keeps the packing elements clean. Moving bed scrubbers are used for particle collection and gas absorption when both processes must be carried out simultaneously. Particle collection efficiency can be good down to particle sizes of one micron. Gas absorption and particulate collection are both enhanced when several moving bed stages are used in series. Pressure drops range from 2.5 to 6 inches water gauge per stage. b. High energy scrubbers. High energy scrubbers utilize high gas velocities to promote removal of particles down to sub-micron size. Gas absorption efficiencies are not very good because of the co-current movements of gas and liquid and resulting limited gas/ liquid contact time.

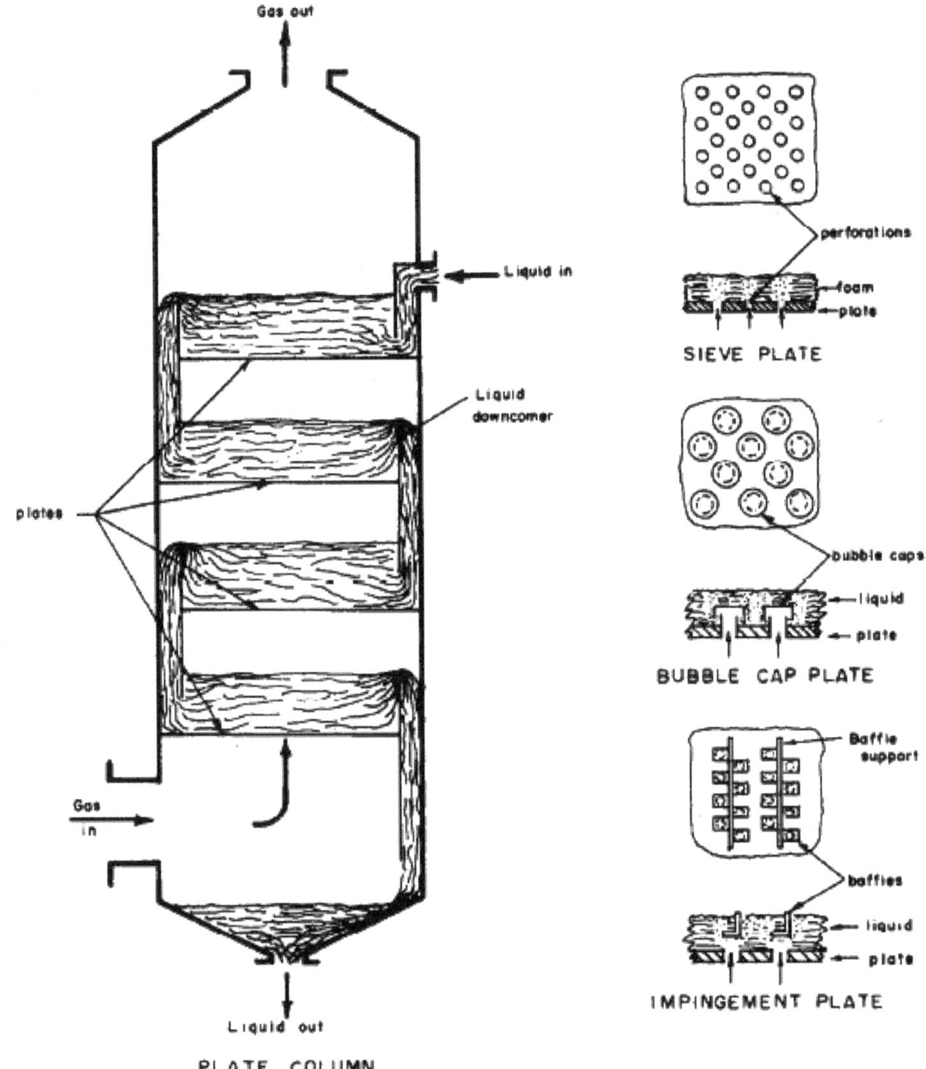

Figure 7-1

Plate type scrubber

© J. Paul Guyer   2021

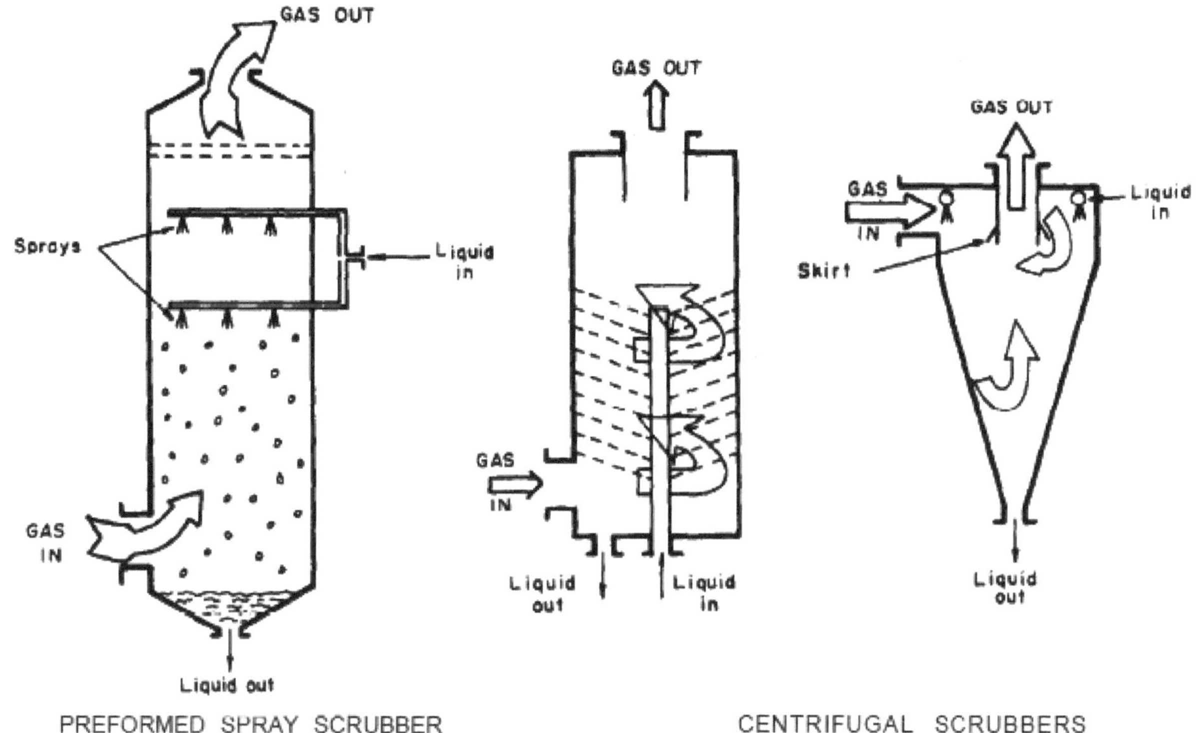

Figure 7-2.

Types of spray and centrifugal scrubbers

3.2.4.1 VENTURI SCRUBBERS. The venturi scrubber utilizes a moving gas stream to atomize and accelerate the liquid droplets (figure 7-4). A convergent- divergent nozzle is used to achieve a gas velocity of 200 to 600 feet per second (ft/ sec) which enhances liquid atomization and particulate capture. Collection efficiency in a gas atomized venturi scrubber increases with pressure drop. Pressure drops of 25 inches water gauge or higher are utilized to collect sub-micron particles. Scrubbers of the gas atomized type have the advantage of adjustment of pressure drop and collection efficiency by varying gas velocity. The gas velocity is controlled by adjusting the area of the venturi throat. Several possible methods for doing this are illustrated in figure 7-5. This can be used to control performance under varying gas flow rates by maintaining a constant pressure drop across the venturi throat. Due to the absence of moving parts, scrubbers of this type may be especially suitable for the collection of sticky particles. Disadvantages include high pressure drop for the collection of sub-micron particles and limited applicability for gas absorption.

3.2.4.2 EJECTOR VENTURI. The ejector venturi scrubber utilizes a high pressure spray to collect particles and move the gas. High relative velocity between drops and gas aids in particle collection. Particle collection efficiency is good for particles larger than a micron in diameter. Gas absorption efficiency is low because of the cocurrent nature of the gas and liquid flow. Liquid pumping power requirements are high and capacity is low making this type impractical for boiler or incinerator emissions control.

3.2.4.3 DYNAMIC (WETTED FAN) SCRUBBER. This scrubber combines a preformed spray, packed bed or centrifugal scrubber with an integral fan to move the gas stream through the scrubber. Liquid is also sprayed into the fan inlet where the rotor shears the liquid into dispersed droplets. The turbulence in the fan increases liquid/ gas contact. This type of scrubber is effective in collection of fine particulate. Construction of this scrubber is more complex due to the necessity of the fan operating in a wet and possibly corrosive gas stream. The design must prevent build-up of particulates on the fan rotor.

3.2.4.4 DRY SCRUBBERS. Dry scrubbers are so named because the collected gas contaminants are in a dry form.

3.2.4.4.1 SPRAY DRYER The spray dryer is used to remove gaseous contaminants, particularly sulfur oxides from the gas stream. An alkaline reagent slurry is mechanically atomized in the gas stream. The sulfur oxides react with the slurry droplets and are absorbed into the droplets. At the same time, the heat in the gas stream evaporates the water from the droplets leaving a dry powder. The gas stream is then passed through a fabric filter or electrostatic precipitator where the dry product and any fly ash particulate is removed. The scrubber chamber is an open vessel with no internals other than the mechanical slurry atomizer nozzles. The vessel is large enough to allow complete drying of the spray before impinging on the walls and to allow enough residence time for the chemical reaction to go to completion. A schematic of the system is shown in figure 7-6.

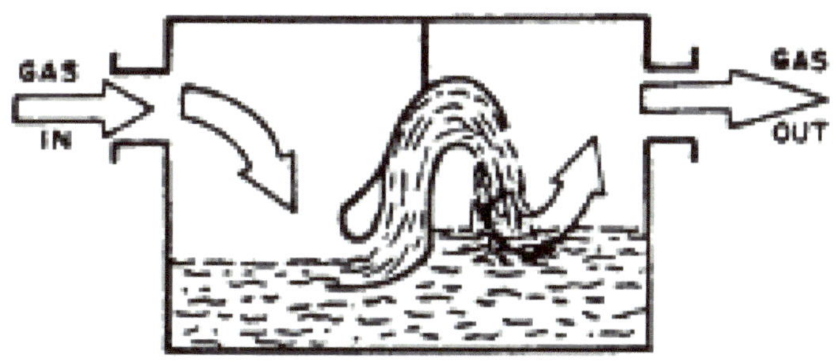

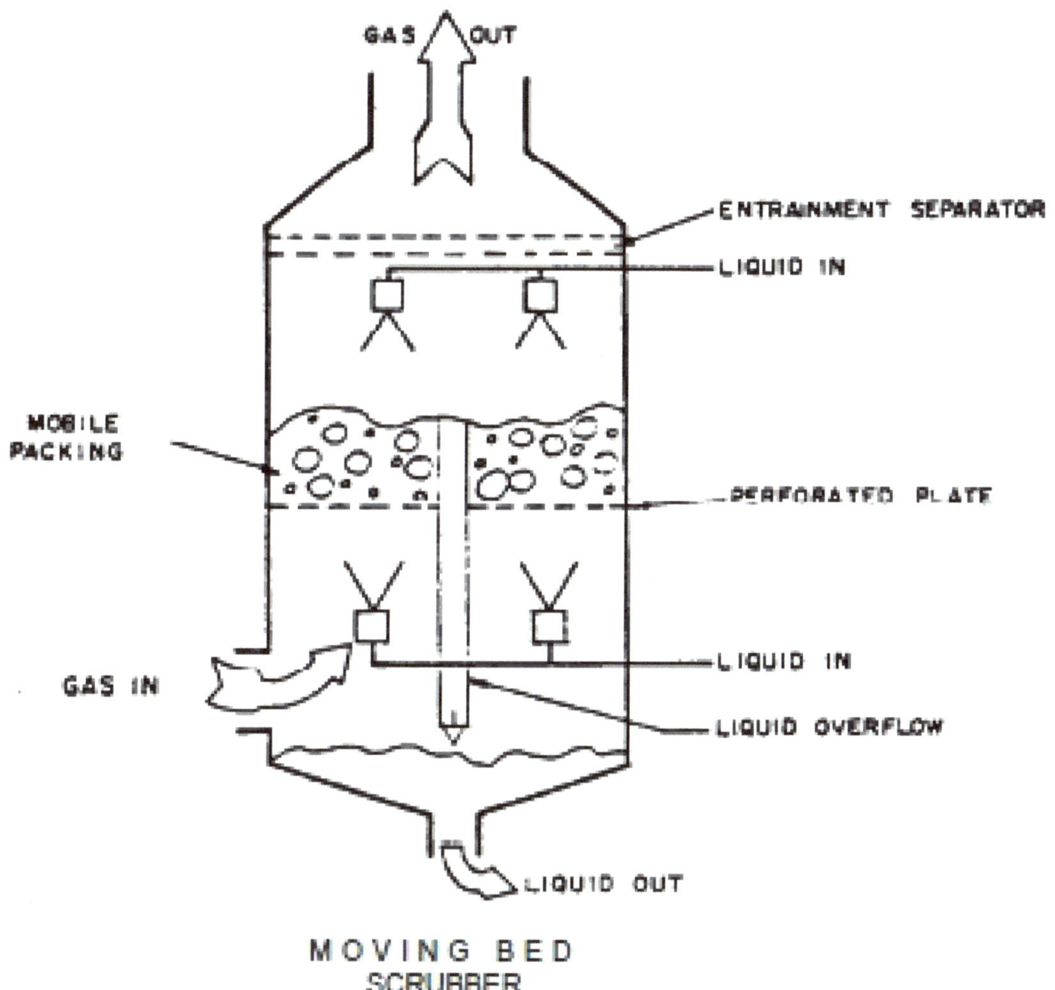

Figure 7-3

Types of entrainment and mowing bed scrubbers

3.2.4.4.2 GRAVEL BED. The gravel bed, while referred to as a dry scrubber, is more a filter using sized gravel as the filter media. A bed of gravel is contained in a vertical cylinder between two slotted screens. As the gas passes through the interstices of the gravel, particulates impact on, and are collected on the gravel surface. Sub-micron size particles are also collected on the surface because of their Brownian movement. Dust-laden gravel is drawn off the bottom and the dust is separated from the gravel by a mechanical vibrator or pneumatic separator. The cleaned gravel is then conveyed up and dumped on top of the gravel bed. The cylindrical bed slowly moves down and is constantly recycled.

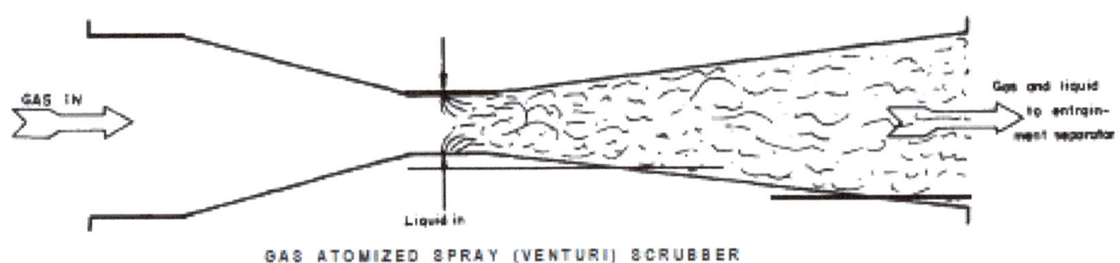

Figure 7-4

Gas atomized spray (venturi) scrubber

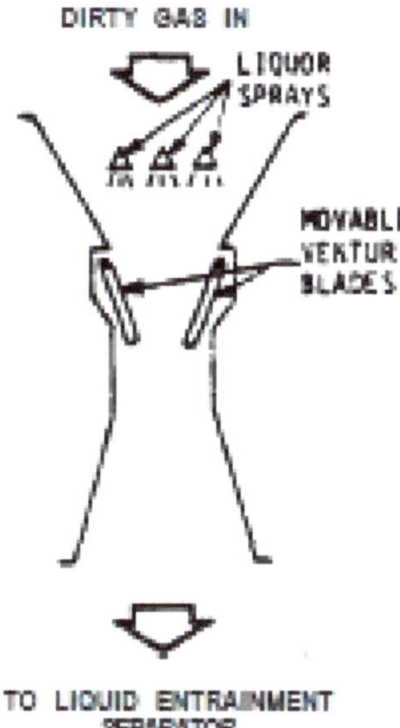

a. Movable-blade venturi

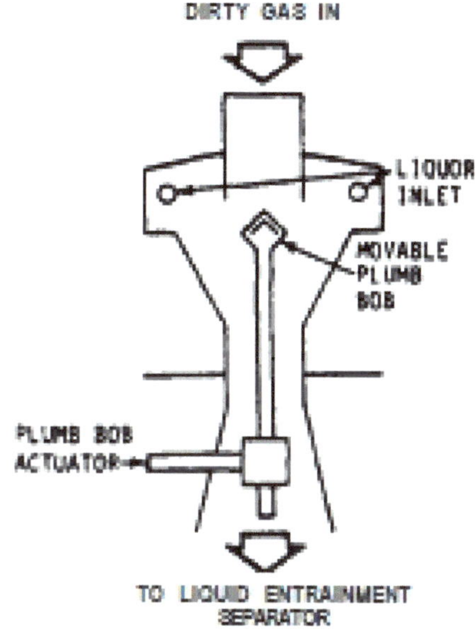

b. Plumb-bob venturi

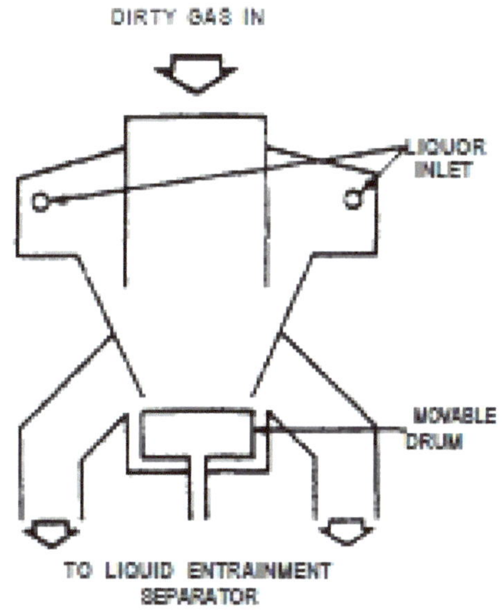

c. Radial-flow venturi

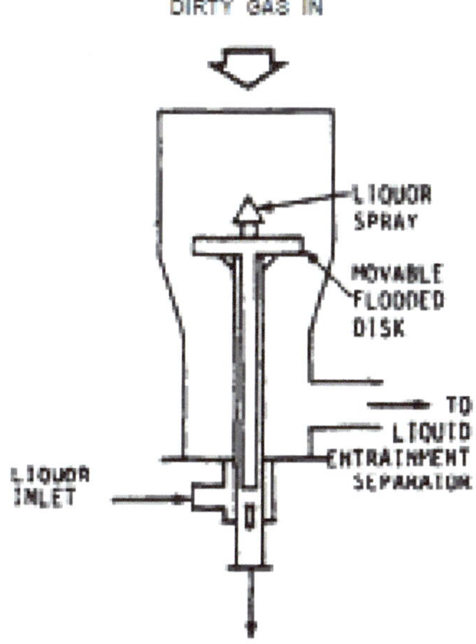

d. Flooded-disc venturi

Figure 7-5

Throat sections of variable venturi scrubbers

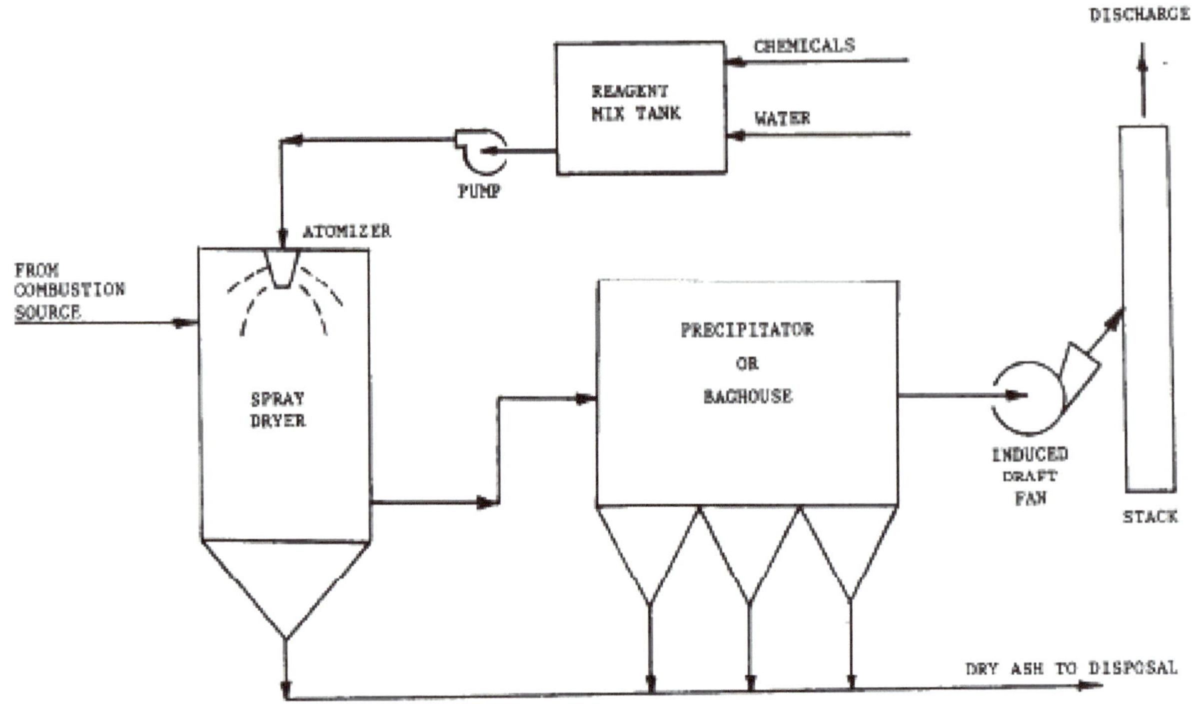

Figure 7-6

Spray dryer system

3.3 APPLICATION

3.3.1 PARTICULATE REMOVAL. Scrubbers may be used as control devices on incinerators and boilers for fly ash collection. The plate, spray, venturi, and moving bed types have been successfully applied; however, their application has been limited because they require:

- -more energy than dry particulate collection devices of the same collection efficiency,
- -water supply and recovery system,
- -more extensive solid waste disposal system,
- -system to control the scrubbing process in response to gas flow rate changes.

3.3.2 IN MAKING DECISIONS ON applicability to a particular process, figure 7-7 is useful in determining all components which must be taken into consideration.

3.3.3 GASEOUS REMOVAL. Scrubbers have been used primarily for the removal of sulfur oxides in stack gases. (See chapter 10 for a more detailed description of sulfur oxides (SOx) control techniques.) However, as new control systems are devised, simultaneous removal of gases and particulate material will become the accepted procedure for designing scrubbers for combustion processes.

3.4 TREATMENT AND DISPOSAL OF WASTE MATERIALS. Wet scrubber systems are designed to process exhaust streams by transfer of pollutants to some liquid medium, usually water seeded with the appropriate reactants. Liquid effluent treatment and disposal are therefore an essential part of every wet scrubber system. Installation and maintenance of the associated components can add appreciably to the system capital and operating costs. The degree of treatment required will depend upon the methods of disposal or recycle and on existing regulations. Required effluent quality, environmental constraints, and availability of disposal sites must be established before design of a treatment facility or the determination of a disposal technique can proceed. In many industrial applications the scrubber liquid wastes are combined with other plant wastes for treatment in a central facility. Design of this waste treatment should be by an engineer experienced in industrial waste treatment and disposal.

3.5 SELECTION OF MATERIALS

3.5.1 GENERAL CONDITIONS. When choosing construction materials for scrubber systems, certain pertinent operating parameters should be considered. The metal surface of an exhaust gas or pollution control system will behave very differently in the same acid mist environment, depending on conditions of carrier gas velocity, temperature, whether the conditions are reducing or oxidizing, and upon the presence of impurities. For example, the presence of ferric or cupric iron traces in acids can dramatically reduce corrosion rates of stainless steels and titanium alloys. On the other hand, traces of chloride or fluoride in sulfuric acid can cause severe pitting in stainless steels. This condition is frequently encountered in an incinerator which burns large quantities of disposable polyvinyl chloride (PVC) materials

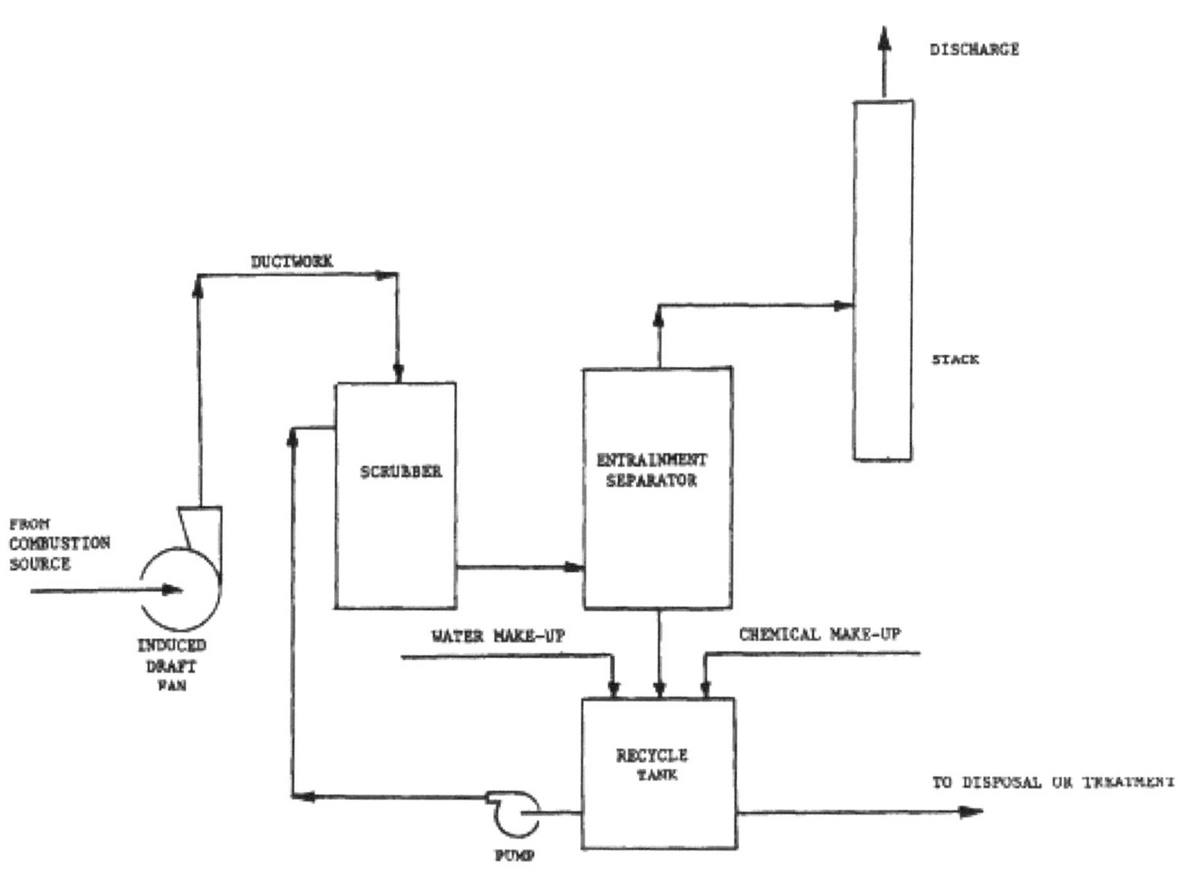

Figure 7-7

Schematic diagram of scrubber flow

© J. Paul Guyer   2021

3.5.2 TEMPERATURE. Corrosion rates generally increase with increases in exhaust temperatures. This is due to the increased mobility of ions and increased reaction rates. However, in cases where the corrosion process is accelerated by the presence of oxygen, increasing the acid temperature eventually boils out dissolved oxygen, rapidly diminishing corrosion rate. This is the case with Monel, a nickel-copper alloy.

3.5.3 VELOCITY. Often the corrosion resistance of an alloy depends on the existence of an adhering oxide layer on its surface. A high exhaust gas velocity can remove or erode the surface layer. Once removed, this layer cannot be renewed because the oxide film is washed away as it forms.

3.5.4 STATE OF OXIDATION. Under reducing condition, Monel is very resistant to moderate sulfuric-acid concentrations. Under oxidizing conditions, or in the presence of oxidizing ions, however, very rapid corrosion occurs. The reverse is true of stainless steels which are resistant to oxidizing acid environments, but are attached by acids under reducing conditions. The equipment designer should select materials based on individual case conditions including temperature, abrasion, pH, etc.

3.6 AUXILIARY EQUIPMENT

3.6.1 GAS TRANSPORT.

3.6.1.1 DUCTS AND STACKS. Large boiler plant stacks have a wind shield of reinforced concrete or of steel, with a separate inner flue or numerous flues of steel, acid-resistant brick, and occasionally, stainless steel. The space between the inner flue and the outer wind shield may be insulated with a mineral wool wrapping. This is to prevent the condensation of acid dew on the inside of the metal chimney, which occurs below dew point temperature, and also to prevent acid "smut" from being blown out of the chimney. Acid smut is a term for ash particles contaminated with acid. It is heavy and tends to fall out of the gas plume soon after exiting from the stack. In smaller plants, stacks may be a single wall steel construction with insulation and lagging on the outer surface. For wet scrubbing practice, chimneys for vapor-saturated gases containing corrosive substances may be made of rubber-lined steel, fiberglass-reinforced resin or other corrosion-resistant material. With materials that have a limited maximum temperature, provisions must be made to protect the stack from high temperatures because of loss of scrubbing liquid. Chimney or stack velocities are generally 30 ft/ sec to prevent re-entrainment of moisture from the stack wall which would rain down around the plant. Sometimes cones are fitted at the top to give exit velocities as high as 75 ft/sec. The chief reason for high velocities is to eject the gases well away from the top of the stack to increase the effective height and to avoid downwash. Downwash can damage the metal structure supporting the stack, the stack itself, or the outside steel of a lined metal stack.

3.6.1.2 FANS. In a wet scrubber system the preferred location for the boiler or incinerator induced draft fan is upstream of the scrubber. This eliminates the need for special corrosion-resistant construction required to handle the wet downstream gas. The fan should be selected to resist build-up of dry ash or erosion of the rotor surfaces. For high dust load applications a radial blade or radial tip blade fan is more durable. In a dry scrubber application the fan should be downstream of the scrubber in the clean gas stream. Here a more efficient air-foil or squirrel-cage rotor can be used.

3.6.2 LIQUID TRANSPORT.

3.6.2.1 PIPEWORK. For most scrubbing duties, the liquid to be conveyed will be corrosive. There exists a wide variety of acid resistant pipework to choose from, but generally speaking, rubber- lined steel pipe has high versatility. It is easy to support, has the strength of steel, will withstand increases in temperature for a short time and will not disintegrate from vibration or liquid hammer. Fiberglass filament wound plastic pipe is also suitable for a very wide range of conditions of temperature, pressure, and chemicals. The chief disadvantage of rubber- lined pipe is that it cannot be cut to size and has to be precisely manufactured with correct lengths and flange drilling. Site fabrication is not possible. Most piping is manufactured to ANSI specifications for pressure piping. Considerations must also be made for weatherproofing against freezing conditions.

3.6.2.2 PUMPS. Centrifugal pumps are used to supply the scrubbing liquid or recycled slurry to the scrubber nozzles at the required volume flow rate and pressure. Where no solids are present in the liquid, bare metal pumps, either iron or stainless steel construction, are used. In recycle systems with solids in the liquid, special rubber-lined or hard-iron alloy pumps are used to control erosion of the pump internals. These are generally belt driven to allow selection of the proper speed necessary for the design capacity and head. Solids content must still be controlled to limit the maximum slurry consistency to meet the scrubber and pump requirements.

3.6.2.3 ENTRAINMENT SEPARATION. After the wetted gas stream leaves the scrubbing section, entrained liquid droplets must be removed. Otherwise they would rain out of the stack and fall on the surrounding area. Removal can be by gravity separation in an expanded vessel with lowered velocity or a cyclonic separator can swirl out the droplets against the vessel wall. Knitted wire or plastic mesh demisters or chevron or "zig-zag" vanes can be located at the scrubber outlet to catch any droplets.

3.6.2.4 PROCESS MEASUREMENT AND CONTROL. The scrubber control system should be designed to follow variations in the boiler or incinerator gas flow and contaminant load to maintain outlet emissions in compliance with selected criteria.

3.6.2.4.1 MEASUREMENTS. Measurement of data from the process to provide proper control should include inlet gas flow rate, temperature and pressure, scrubber gas pressure drop, liquid pressure, flow rate, solids consistency, pH, and outlet gas temperature. Selection of instrumentation hardware should be on an individual application basis.

3.6.2.4.2 CONTROL. Pressure drop across a scrubber can be referenced as an indication of performance following initial or periodic, outlet gas testing. In a variable throat venturi, for instance, this pressure drop can be used to control the throat opening, maintaining constant performance under varying gas volume flow rates. Measurement of scrubber slurry solids consistency can be used to control bleed-off of high solids slurry and make-up with fresh water. If sulfur dioxide (SO2) is being controlled then measurement of scrubber liquid pH can control make-up of caustic to maintain efficiency of SO2 removal. Complete specification or design of a control system must be on a case-by-case basis.

3.7 ADVANTAGES AND DISADVANTAGES.

3.7.1 ADVANTAGES. The advantages of selecting scrubbers over other collection devices are:

- -Capability of gas absorption for removal of harmful and dangerous gases,
- -High efficiency of particulate removal,
- -Capability of quenching high temperature exhaust gases,
- -Capability of controlling heavy particulate loadings,

3.7.2 DISADVANTAGES. The disadvantages of selecting scrubbers over other collection devices are:

- -Large energy usage for high collection efficiency,
- -High maintenance costs,
- -Continuous expenses for chemicals to remove gaseous materials,
- -Water supply and disposal requirements,
- -Exhaust gas reheat may be necessary to maintain plume dispersion,
- -Weather proofing is necessary to prevent freezeup of equipment.

3.8 ELECTROSTATIC PRECIPITATOR

3.8.1 GENERAL. An electrostatic precipitator (ESP) is a device which removes particles from a gas stream. It accomplishes particle separation by the use of an electric field which:

- -imparts a positive or negative charge to the particle,
- -attracts the particle to an oppositely charged plate or tube,
- -removes the particle from the collection surface to a hopper by vibrating or rapping the collection surface.

3.8.2 TYPES OF ELECTROSTATIC PRECIPITATORS.

3.8.2.1 TWO STAGE ESPs. Two stage ESPs are designed so that the charging field and the collecting field are independent of each other. The charging electrode is located upstream of the collecting plates. Two stage ESPs are used in the collection of fine mists.

3.8.2.2 SINGLE STAGE ESPs. Single stage ESPs are designed so that the same electric field is used for charging and collecting particulates. Single stage ESPs are the most common type used for the control of particulate emissions and are either of tube or parallel plate type construction. A schematic view of the tube and parallel plate arrangement is given in figure 8-1. The tube type precipitator is a pipe with a discharge wire running axially through it. Gas flows up through the pipe and collected particulate is discharged from the bottom. This type of precipitator is mainly used to handle small gas volumes. It possesses a collection efficiency comparable to the parallel plate types, usually greater than 90 percent. Water washing is frequently used instead of rapping to clean the collecting surface. Parallel plate precipitators are the most commonly used precipitator type. The plates are usually less than twelve inches apart with the charging electrode suspended vertically between each plate. Gas flow is horizontal through the plates.

3.8.3 MODES OF OPERATION. All types of ESPs can be operated at high or low temperatures, with or without water washing (table 8-I).

3.8.3.1 HOT PRECIPITATION. A hot precipitator is designed to operate at gas temperatures above 600 degrees Fahrenheit and is usually of the single stage, parallel plate design. It has the advantage of collecting more particulate from the hot gas stream because particle resistance to collection decreases at higher temperatures. The ability to remove particles from the collection plates and hoppers is also increased at these temperatures. However, hot precipitators must be large in construction in order to accommodate the higher specific volume of the gas stream.

3.8.3.2 COLD PRECIPITATION. Cold precipitators are designed to operate at temperatures around 300 degrees Fahrenheit. The term "cold" is applied to any device on the low temperature side of the exhaust gas heat exchanger. Cold ESPs are also generally of the single stage, parallel plate design. They are smaller in construction than hot precipitator types because they handle smaller gas volumes due to the reduced temperature. Cold precipitators are most effective at collecting particles of low resistivity since particle resistance to collection is greater at lower temperatures. These precipitators are subject to corrosion due to the condensation of acid mist at the lower temperatures.

3.8.3.3 WET PRECIPITATION. A wet precipitator uses water to aid in cleaning the particulate collection plates. It may employ water spray nozzles directed at the collection plates, or inject a fine water mist into the gas stream entering the precipitator. Wet precipitators enhance the collection efficiency of particulates by reducing reentrainment from the collection plates. Care should be taken so that water addition does not lower gas temperature below the dewpoint temperature, thus allowing the formation of acids. A wet precipitator can be of either plate or tube type construction.

3.8.4 APPLICATIONS. Electrostatic precipitators are among the most widely used particulate control devices. They are used to control particulate emissions from the electric utility industry, industrial boiler plants, municipal incinerators, the non-ferrous, iron and steel, chemical, cement, and paper industries. It is outside the scope of this manual to include all of these application areas. Only applications to boilers and incinerators will be reviewed.

3.8.4.1 BOILER APPLICATION. Parallel plate electrostatic precipitators are commonly employed in the utility industry to control emissions from coal-fired boilers. Cold type precipitators are the prevalent type because they are most easily retrofitted. In the design of new installations, the use of hot precipitators has become more common, because of the greater use of lower sulfur fuels. Low sulfur fuels have higher particle resistivity and therefore particulate emissions are more difficult to control with cold precipitation. Figure 8-2 may be used for estimating whether hot precipitators or cold precipitators should be selected for a particular sulfur content of coal.

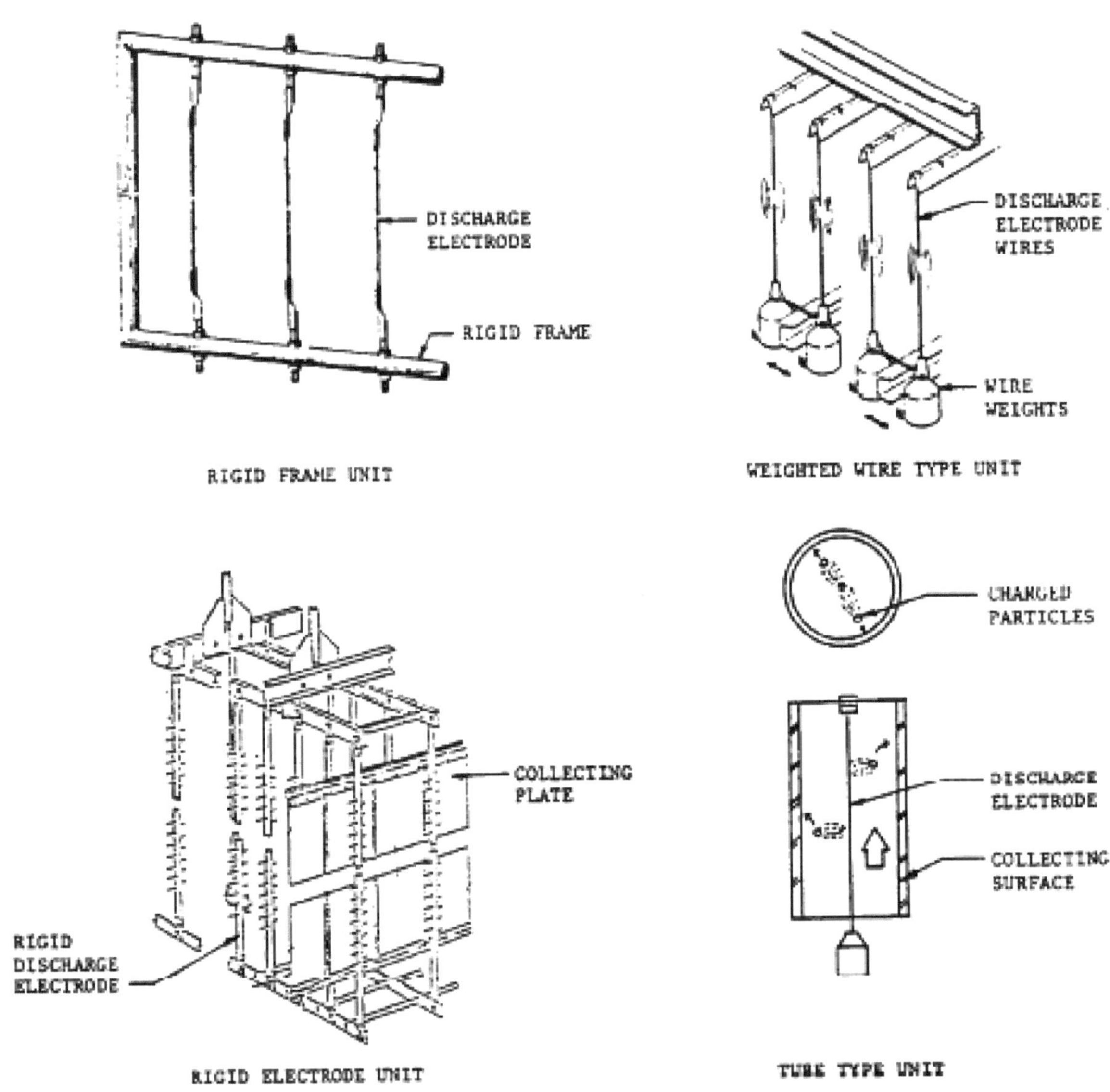

Figure 8-4

Schematic views of flat end tubular surface type electrostatic precipitators

3.8.4.2 WOOD REFUSE BOILER APPLICATIONS. An ESP can be used for particulate collection on a wood fired boiler installation if precautions are taken for fire prevention.

The ESP should be preceded by some type of mechanical collection device to prevent hot glowing char from entering the precipitator and possibly starting a fire.

3.8.4.3 INCINERATOR APPLICATION. Until relatively recently, ESPs were used for pollution control on incineration units only in Europe. In the United States, however, the ESP is now being viewed as one of the more effective methods for the control of emissions from incinerators. The major problem associated with the use of precipitators on incinerators is high gas temperatures. Temperatures up to 1800 degrees Fahrenheit can be encountered at the incinerator outlet. These temperatures must be reduced before entering a precipitator. Several methods can be used to accomplish this temperature reduction:

- -mixing of the gas with cooler air,
- -indirect cooling such as waste heat boilers,
- --evaporative cooling in which droplets of water are sprayed into the gas.

3.8.5 PERFORMANCE. The performance of an electrostatic precipitator is predominantly affected by particle resistivity, particle size, gas velocity, flow turbulence, and the number of energized bus sections (electrically independent sections) in operation.

3.8.5.1 PARTICLE RESISTIVITY. Particle resistivity is an electrical property of a particle and is a measure of its resistance of being collected. Particle resistivity is affected by gas temperature, humidity, sodium content, and sulfur trioxide (SO_3) content. See figure 8-3.

3.8.5.2 COLLECTION PLATE AREA. Collection plate area, and gas volume, affect electrostatic precipitator performance. The basic function relating these factors is shown in equation 8-I.

Type	Operating Temperature °F	Dust Resistivity at 300°F ohm-cm	Gas Flow ft^3/min.	Pressure Drop in. of water	Design Collection Efficiency % by weight	Application	Other
Hot ESP	600+	greater than 10^{12}	100,000+	less than 1"	Usually 90+ can go to 99+	Before pre-heater in boilers, incinerator, industrial	Can collect high resistivity dust. Has higher gas flow and is large in size. Corrosion usually not a problem.
Cold ESP	300	less than 10^{10}	100,000+	less than 1"	Usually 90+ Can go to 99+	After pre-heater in boilers, incinerator, industrial	Limited to dust resistivities lower than 10^{10} ohm-cm. Corrosion can be a problem.
Wet ESP	300-	greater than 10^{12} below 10^4	100,000	less than 1"	Usually 90+ Can go to 99+	Industrial, boilers, incinerator	Useful for high or low resistivity dust collection. Corrosion usually not a problem.

TABLE 8-1

OPERATING CHARACTERISTICS OF PRECIPITATORS

© J. Paul Guyer   2021

$$CE = 1 - e^{-\left(\dfrac{A_c}{V_g} \times w\right)^p} \qquad \text{(eq. 8-1)}$$

Where: CE = collection efficiency
A_c = collection plate area in square feet (ft^2)
V_g = gas flow rate in cubic feet/minute (ft^3/min)
w = migration velocity or precipitation rate parameter, feet/minute (ft/min).

3.8.5.3 BUS SECTIONS. The number of energized bus sections in a precipitator has an effect upon collection efficiency. A power loss in one energized bus section will reduce the effectiveness of the precipitator. See figure 8-4.

3.8.5.4 TURBULENCE. Turbulence in the gas flow through an electrostatic precipitator will decrease its collection efficiency. For proper operation all segments of the flow should be within 25 percent of the mean flow velocity.

3.8.6 DESCRIPTION OF COMPONENTS

3.8.6.1 SHELL. The shell of an ESP has three main functions: structural support, gas flow containment, and insulation. Shell material is most commonly steel; if necessary, insulation can be applied to the exterior to prevent heat loss, Brick or concrete linings can be installed on shell interiors if gas stream corrosion of the metal may occur. Corrosion resistant steel can also be used as a lining, but the cost may be uneconomical and at times prohibitive. Since the shell is also used for structural support, normal civil engineering precautions should be taken in the design.

3.8.6.2 WEIGHTED WIRE DISCHARGE ELECTRODES. Wires vary in type, size, and style. Provision is made to keep the discharge wire from displacement by attachment to a suspended weight. The wires can be made stiff, consisting of a formed sheet, or they can be simple variations of the normal straight round wire such as being barbed or

© J. Paul Guyer   2021

pronged. Steel alloys are commonly used for wire construction, but actually any
conducting material with a proper configuration and sufficient tensile strength can be
used.

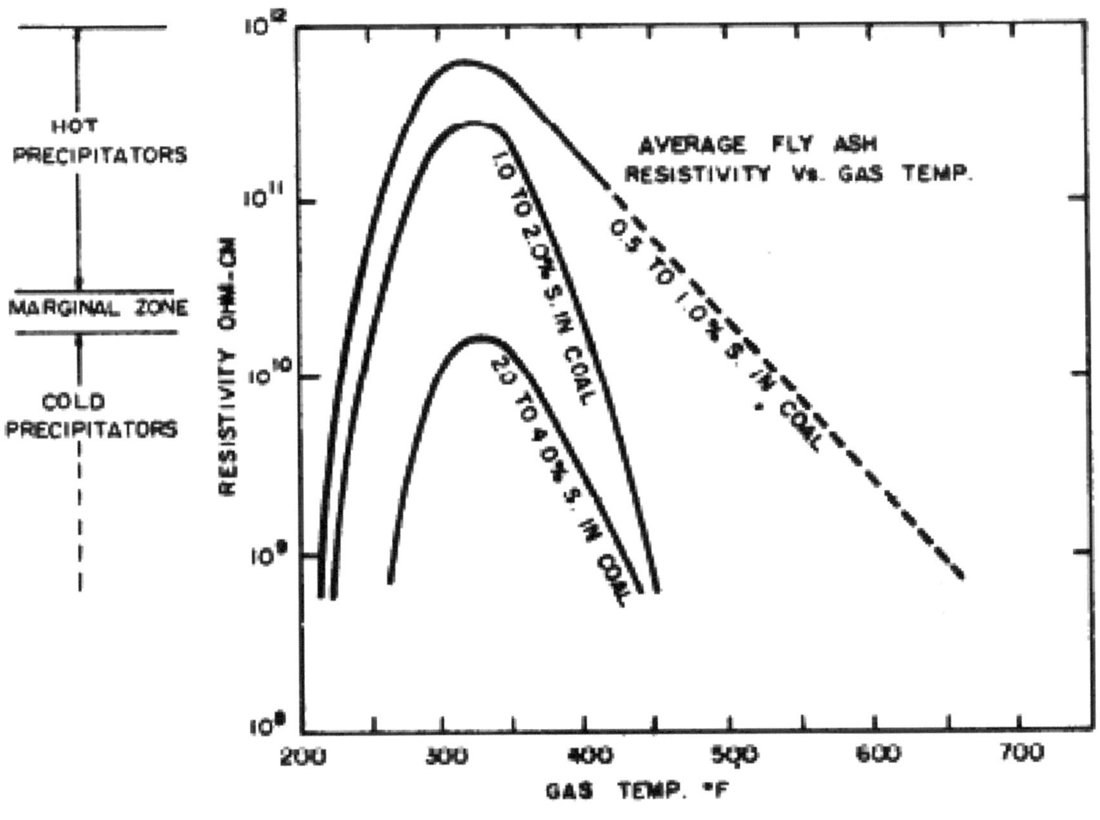

COAL FLY ASH

RESISTIVITY-SULPHUR-TEMPERATURE

Figure 8-2

Operating ranges for hot/cold electrostatic precipitators

3.8.6.2.1 RIGID FRAME DISCHARGE ELECTRODES. Rigid frame designs incorporate
a framework which supports the discharge electrodes. By using the rigid frame design
the need for wire weights is eliminated since the frame keeps the wires properly
supported and aligned.

3.8.6.2.2 RIGID ELECTRODES. The rigid electrode design uses electrodes that have sufficient strength to stay in alignment their entire length. The electrodes are supported from the top and kept in alignment by guides at the bottom. Rigid electrodes are the least susceptible to breakage.

3.8.6.2.3 COLLECTION ELECTRODES. There are numerous types of collection electrodes designed to minimize reentrainment and prevent sparking. The material used in construction, however, must be strong enough to withstand frequent rapping. In order to insure correct electrode application, it is wise to see if the electrode chosen has exhibited good performance at similar installations.
3.

8.6.2.4 HOPPERS. A hopper is used to collect ash as it falls from the precipitator. The hopper should be designed using precautions against corrosion in the precipitator, as any leakage due to corrosion will enhance entrainment. If the precipitator is dry, a hopper angle should be chosen that will prevent bridging of collected dust.

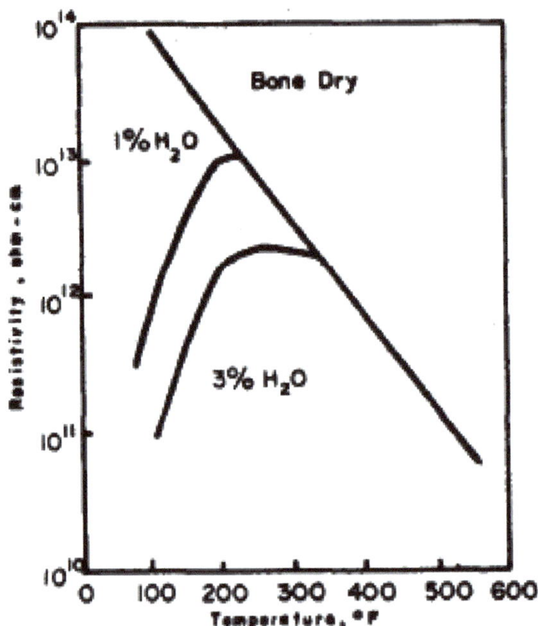

Effect of gas humidity in increasing
the conductivity of a typical fly ash.

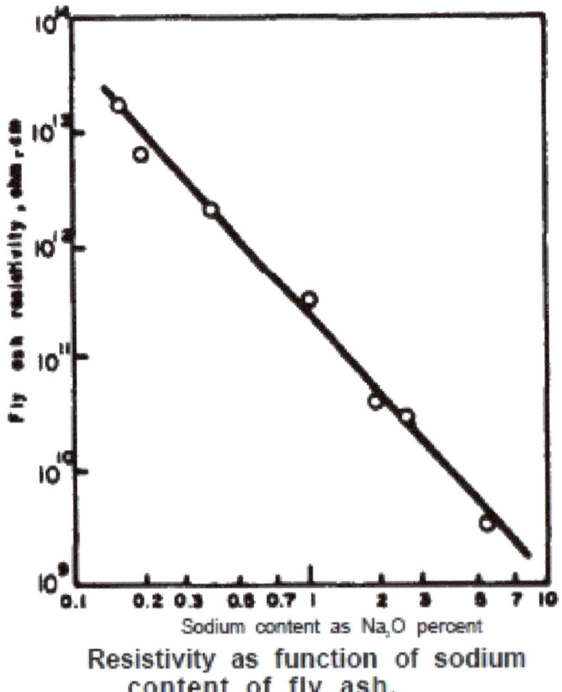

Resistivity as function of sodium
content of fly ash.

Figure 8-3

Factors affecting particle resistivity

© J. Paul Guyer   2021

Hoppers must be sized so that the amount of dust collected over a period of time is not great enough to overflow and be re-entrained. Seals also must be provided around the outlet to prevent any air leakage. If the precipitator is wet, the hopper should allow removal of sludge in a manner compatible with the overall removal system. In general the collected dust in the hoppers is more free-flowing when kept hot. The hoppers should be insulated and should have heaters to maintain the desired temperatures. Hoppers heaters will also prevent the formation of acids that may occur at low temperatures. Provisions should be made for safe rodding out the hoppers should they become plugged.

3.8.6.2.5 RAPPERS. Rappers are used to remove dust from the discharge and collection electrodes. Rappers are usually one of two types, impulse or vibrator. The vibrator type removes dust from the discharge electrode by imparting to it a continuous vibration energy. They are used to remove dust from the collection electrodes. Impulse rappers consist of electromagnetic solenoids, motor driven cams, and motor driven hammers. Important features to note in choosing rappers are long service life without excessive wear and flexible enough operation to allow for changing precipitator operating conditions. Low intensity rapping of plates (on the order of one impact per minute) should be used whenever possible to avoid damage to the plates. Visual inspection of the effect of rapping on reentrainment is usually sufficient to determine a good rapping cycle.

3.8.6.2.6 HIGH TENSION INSULATORS. High tension insulators serve both to support the discharge electrode frame and also to provide high voltage insulation. The materials used are ceramic, porcelain, fused silica and alumina. Alumina is the most common. The insulators must be kept clean to prevent high voltage shorting and resultant equipment damage. Compressed air or steam can be used for this purpose.

3.8.6.2.7 FOUR POINT SUSPENSION. Rigid electrode and rigid frame units may utilize a four point suspension system to support the discharge electrode framework in each chamber. This type of suspension system assures a better alignment of the discharge and collection electrodes. This in turn provides a more consistent operation.

3.8.6.2.8 DISTRIBUTION DEVICES. Perforated plates, baffles or turning vanes are usually employed on the inlet and outlet of an ESP to improve gas distribution. Improper distribution can cause both performance and corrosion problems. These distribution devices may require rappers for cleaning.

3.8.6.2.9 MODEL TESTING. Gas flow models are used to determine the location and type of distribution devices. The models may include both the inlet and outlet ductwork in order to correctly model the gas flow characteristics. Gas flow studies may not be required if a proven precipitator design is installed with a proven ductwork arrangement.

3.8.7 CONTROL SYSTEMS. The electric power control system is the most important component system of any ESP. The basic components of this system are: step-up transformer; high voltage rectifier; voltage and amperage controls; and sensors.

3.8.7.1 AUTOMATIC POWER CONTROL. By utilizing a signal from a stack transmission meter the power level in the precipitator can be varied to obtain the desired performance over a wide range of operating conditions.

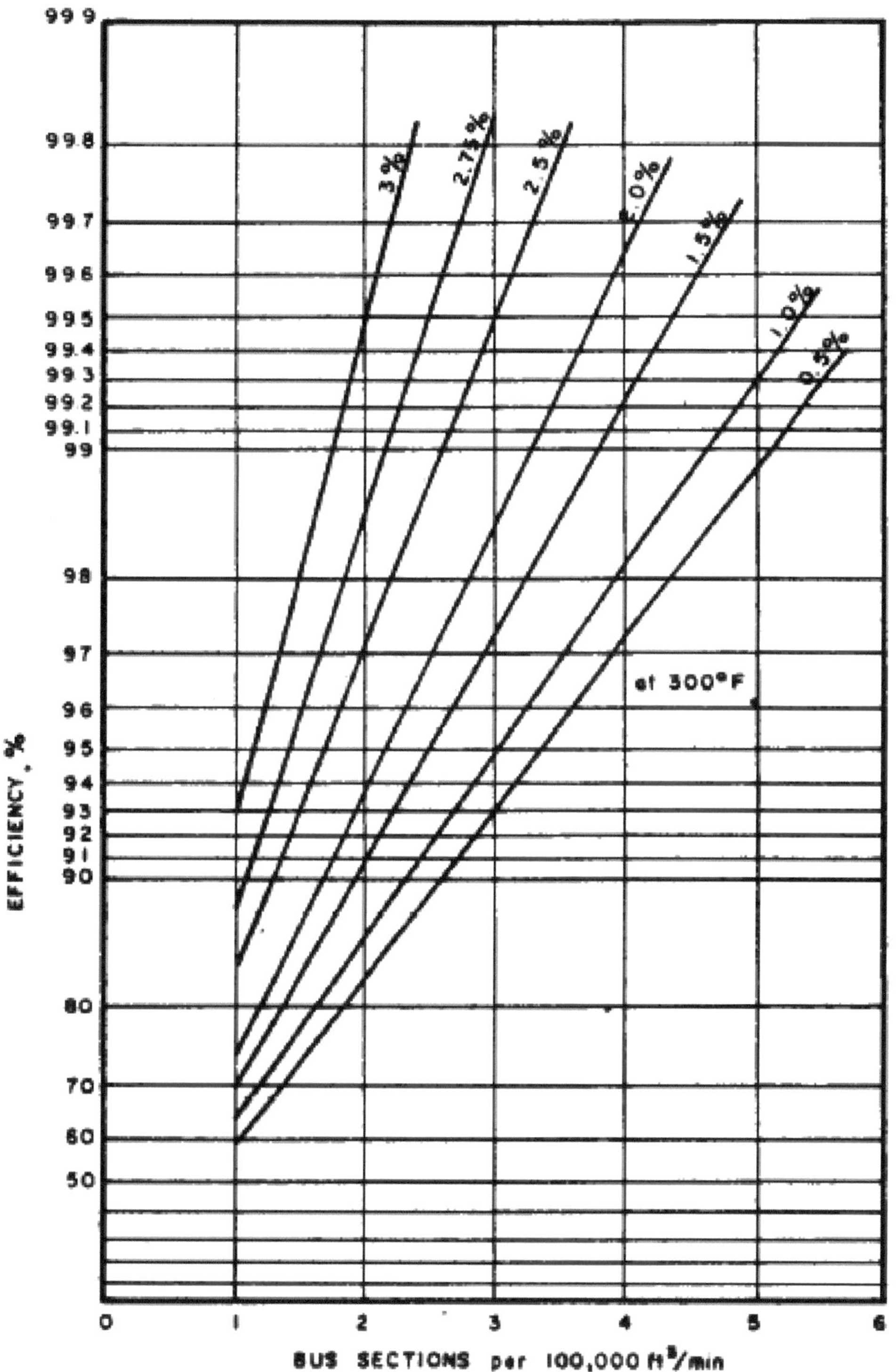

Figure 8-4

Bus sections vs. efficiency for different sulfur percentages in coal.

3.8.7.2 HIGH VOLTAGE TRANSFORMER. The standard iron core transformer is the only instrument generally used to step-up the input voltage. The only care that need be taken is that the transformer is of superior quality and able to put out the quantity of voltage required by the precipitator. Transformers are designed to withstand high ambient temperatures and electrical variations induced by sparking. For high temperature operation, the most common transformer cooling method is liquid immersion.

3.8.7.3 HIGH VOLTAGE RECTIFIER. Silicon rectifiers are the latest advance in rectifying circuitry. They are solid state devices which have a few of the disadvantages of the other types of rectifiers. An assembly of silicon rectifiers is used for lower rated current sets, typically 500 mili-amperes (mA).

3.8.7.4 VOLTAGE AND AMPERAGE CONTROLS. Controls are needed to insure that the precipitator is supplied with the maximum amount of voltage or power input, and to control the effects of sparking. The most modern method of accomplishing these aims is through the use of silicon controlled rectifiers (SCR). Other modern control devices are saturable reactors and thyristors (four element, solid state devices). Voltage control can also be accomplished by tapped series dropping resistors, series rheostats, tapped transformer primaries, and variable inductances.

3.8.7.5 AUXILIARY CONTROL EQUIPMENT. As with any control device, gas flow should be monitored either by read-out of amperage from the fans or by measuring static pressure. It is also useful to have sensors which measure the sulfur dioxide (SO_2) concentration and temperature of the inlet gas stream in order to determine the dewpoint temperature.

3.8.8 ADVANTAGES AND DISADVANTAGES

3.8.8.1 ADVANTAGES.

- The pressure drop through a precipitator is a function of inlet and outlet design and precipitator length. Pressure drop rarely exceeds 0.5 inches, water gauge.
- The ESP can be designed to have 99.9 + percent collection efficiency.
- Silicon control rectifiers and other modern control devices allow an electrostatic precipitator to operate automatically.
- Low maintenance costs.

3.8.8.2 DISADVANTAGES.

- Due to the size of a typical ESP and the erratic nature of most processes (especially if frequent start-up and shutdowns occur) the temperature in different parts of the structure could at times drop below the acid dew point. Corrosion can cause structural damage and allow air leakage.
- An ESP is sensitive to its design parameters. A change in the type of coal used, for example, could drastically affect performance.
- High capital costs.
- If particulate emission concentrations are high, a mechanical precleaner may be necessary.
- High voltages are required.
- No SO_2 control is possible with an ESP.

CHAPTER 4

SULFUR AND NITROGEN OXIDES CONTROL

4.1. FORMATION OF SULFUR OXIDES (SOX)

4.1.1 DEFINITION OF SULFUR OXIDE. All fossil fuels contain sulfur compounds, usually less than 8 percent of the fuel content by weight. During combustion, fuel-bound sulfur is converted to sulfur oxides in much the same way as carbon is oxidized to CO2. Sulfur dioxide (SO2) and sulfur trioxide (SO3) are the predominant sulfur oxides formed. See equations 10-l and 10-2.

$$S + O_2 \quad \text{-----}> \quad SO_2 \qquad \text{(eq. 10-1)}$$
$$2\,s\,O_2 + O_2 \quad \text{-----}> \quad 2\,S\,O_3 \qquad \text{(eq. 10-2)}$$

4.1.2 STACK-GAS CONCENTRATIONS. In efficient fuel combustion processes, approximately 95 percent of the fuel-bound sulfur is oxidized to sulfur dioxide with 1 to 2% being cover-ted to sulfur trioxide.

4.1.3 FACTORS AFFECTING THE FORMATION OF SOX.

- SO3 formation increases as flame temperature increases. Above 3,150 degrees Fahrenheit, SO3 formation no longer increases.
- SO3 formation increases as the excess air rate is increased.
- SO3 formation decreases with coarser atomization.

4.2 AVAILABLE METHODS FOR REDUCING SOX EMISSIONS

4.2.1 FUEL SUBSTITUTION. Burning low sulfur fuel is the most direct means of preventing a SOx emissions problem. However, low sulfur fuel reserves are decreasing and are not available in many areas. Because of this, fuel cleaning technology has receive much attention. There are presently more than 500 coal cleaning plants in this country. At present, more than 20% of the coal consumed yearly by the utility industry is cleaned. Forty to ninety percent of the sulfur in coal can be removed by physical cleaning, depending upon the type of sulfur deposits in the coal. As fuel cleaning technology progresses and the costs of cleaning decrease, fuel cleaning will become a long term solution available for reducing sulfur oxide emissions.

4.2.2 CONSIDERATIONS OF FUEL SUBSTITUTION. Fuel substitution may involve choosing a higher quality fuel grade; or it may mean changing to an alternate fuel type. Fuel substitution may require any of the following considerations:

- Alternations in fuel storage, handling, preparation, and combustion equipment.
- When changing fuel type, such as oil to coal, a new system must be installed.
- When choosing a higher quality fuel, as in changing from residual to distillate fuel oil, modest modifications, such as changing burner tips, and oil feed pumps, are required.

4.2.3 CHANGES IN FUEL PROPERTIES. Consideration of possible differences in fuel properties is important. Some examples are:

- Higher ash content increases particulate emissions.
- Lower coal sulfur content decreases ash fusion temperature and enhances boiler tube slagging.
- Lower coal sulfur content increases fly-ash resistivity and adversely affects electrostatic precipitator performance.
- Low sulfur coal types may have higher sodium content which enhances fouling of boiler convection tube surfaces.

- The combination of physical coal cleaning and partial flue gas desulfurization enables many generating stations to meet SO2 standards at less expense than using flue gas desulfurization alone.

4.2.4 MODIFICATION OF FUEL. Some possibilities are:

- Fuels of varying sulfur content may be mixed to adjust the level of sulfur in the fuel to a low enough level to reduce SO2 emissions to an acceptable level.
- Fuels resulting from these processes will become available in the not too distant future. Gasification of coal removes essentially all of the sulfur and liquefaction of coal results in a reduction of more than 85% of the sulfur.

4.2.5 APPLICABILITY OF BOILER CONVERSION FROM ONE FUEL TYPE TO ANOTHER. Table 10-I indicates that most boilers can be converted to other type of firing but that policies of the agencies must also be a consideration.

As designed			Convertible to:		
Coal	Oil	Gas	Coal	Oil	Gas
**X	x	-	-	-	yes
**X	-	X	-	yes	-
**X	-	-	-	yes	yes
-	X	-	*possible	-	yes
-	-	X	*possible	yes	-

Note: *Large DOD boilers must be convertible to coal firing.
\
* * Changing from coal to oil or gas firing is not in accordance with present AR 420-49.

Table 10-1

Convertibility of steam boilers

4.2.6 APPROACH TO FUEL SUBSTITUTION. An approach to fuel substitution should proceed in the following manner:

- Determine the availability of low sulfur fuels.

- For each, determine which would have sulfur emissions allowable under appropriate regulations.
- Determine the effect of each on particulate emissions, boiler capacity and gas temperatures, boiler fouling and slagging, and existing particulate control devices.
- Identify the required equipment modifications, including transport, storage, handling, preparation, combustion, and control equipment.
- For the required heat output calculate the appropriate fuel feed rate.
- Determine fuel costs.
- Determine the cost of boiler and equipment modification in terms of capital investment and operation.
- Annualize fuel costs, capital charges, and operating and maintenance costs.
- With the original fuel as a baseline, compare emissions and costs for alternate fuels.

4.2.7 MODIFICATION TO BOILER OPERATIONS AND MAINTENANCE.

- A method of reducing sulfur oxides emissions is to improve the boiler use of the available heat. If the useful energy release from the boiler per unit of energy input to the boiler can be increased, the total fuel consumption and emissions will also be reduced.
- An improvement in the boiler release of useful energy per unit of energy input can be achieved by increasing boiler steam pressure and temperature. Doubling the steam drum pressure can increase the useful heat release per unit of energy input by seven percent. Increasing the steam temperature from 900 to 1000 degrees Fahrenheit can result in an improvement in the heat release per unit of energy input of about 3.5 percent.
- Another way to maximize the boiler's output per unit of energy input is to increase the attention given to maintenance of the correct fuel to air ratio. Proper automatic controls can perform this function with a high degree of accuracy.
- If additional emphasis can be put on maintenance tasks which directly effect the boilers ability to release more energy per unit of energy input they should be

considered a modification of boiler operations. Items which fall into this category are:

- o   -Washing turbine blades
- o   -adjusting for maximum throttle pressure
- o   -adjusting turbine control valves to insure proper lift
- o   -adjusting preheater seals and feedwater heaters
- o   -insuring cleanliness of heat transfer surfaces, such as condensers, superheaters, reheaters, and air heaters.

4.2.8 LIMESTONE INJECTION. One of the earliest techniques used to reduce sulfur oxide emission was the use of limestone as a fuel additive. This technique involves limestone injection into the boiler with the coal or into the high temperature zone of the furnace. The limestone is calcined by the heat and reacts with the SO_2 in the boiler to form calcium sulfate. The unreacted limestone, and the fly ash are then collected in an electrostatic precipitator, fabric bag filter, or other particulate control device. There are a number of problems associated with this approach:

- The sulfur oxide removal efficiency of the additive approach is in the range of 50 to 70% in field applications. However, it is considered feasible that when combined with coal cleaning, it is possible to achieve an overall SO_2 reduction of 80 percent.
- The limestone used in the process cannot be recovered.
- The addition of limestone increases particulate loadings. In the precipitator this adversely affects collection efficiency.
- The effects of an increased ash load on slagging and fouling as well as on particulate collection equipment present a group of problems which must be carefully considered.
- The high particulate loadings and potential boiler tube fouling in high heat release boilers tend to cause additional expense and technical problems associated with handling large particulate loadings in the collection equipment.
- There have been many claims over the years regarding the applicability of fuel additives to the reduction of sulfur oxide emissions. The United States

Environmental Protection Agency has tested the effect of additives on residual and distillate oil-tired furnaces. They conclude that the additives have little or no effect.

4.2.9 FLUE GAS DESULFURIZATION (FGD). There are a variety of processes which have demonstrated the ability to remove sulfur oxides from exhaust gases. Although this technology has been demonstrated for some time, its reduction to sound engineering practice and widespread acceptance has been slow. This is particularly true from the standpoint of high system reliability. The most promising systems and their performance characteristics are shown in table 10-2.

4.2.10 BOILER INJECTION OF LIMESTONE WITH WET SCRUBBER. In this system limestone is injected into the boiler and is calcined to lime. The lime reacts with the SO_2 present in the combustion gases to form calcium sulfate and calcium sulfite. As the gas passes through a wet scrubber, the limestone, lime, and reacted lime are washed with water to form sulfite. As the gas passes through a wet scrubber, the limestone, lime, and reacted lime are washed with water to form a slurry. The resulting effluent is sent to a settling pond and the sediment is disposed by landfilling. Removal efficiencies are below 50% but can be reliably maintained. Scaling of boiler tube surfaces is a major problem.

4.2.11 SCRUBBER INJECTION OF LIMESTONE. In this FGD system limestone is injected into a scrubber with water to form a slurry (5 to 15% solids by weight). The limestone is ground into fines so that 85% passes through a 200- mesh screen. $CaCO_3$ absorbs SO_2 in the scrubber and in a reaction tank where additional time is allowed to complete the reaction. Makeup is added to the reusable slurry as necessary and the mixture is recirculated to the scrubber. The dischargeable slurry is taken to a thickener where the solids are precipitated and the water is recirculated to the scrubber. Limestone scrubbing is a throwaway process and sludge disposal may be a problem. At SO_2 removal efficiencies of about 30%, performance data indicate that limestone scrubbers have a 90% operational reliability. Plugging of the demister, and corrosion and erosion of stack gas reheat tubes have been major problems in limestone scrubbers. Figure 10-1 shows a simplified process flow-sheet for a typical limestone scrubbing installation.

4.2.12 SCRUBBER INJECTION OF LIME. This FGD process is similar to the limestone scrubber process, except that lime (Ca(OH)2) is used as the absorbent. Lime is a more effective reactant than limestone so that less of it is required for the same SO2 removal efficiency. The decision to use one system over the other is not clear-cut and usually is decided by availability.

4.2.13 POST FURNACE LIMESTONE INJECTION WITH SPRAY DRYING. In this system, a limestone slurry is injected into a spray dryer which receives flue gas directly from the boiler. The limestone in the slurry reacts with the SO2 present in the combustion gases to form calcium sulfate and calcium sulfite. The heat content of the combustion gases drives off the moisture resulting in dry particulates exiting the spray dryer and their subsequent capture in a particulate collector following the spray dryer. The particulates captured in the collector are discharged as a dry material and the cleaned flue gases pass through the filter to the stack (fig lo-la).

4.2.14 DRY, POST FURNACE LIMESTONE INJECTION. Ground dry limestone is injected directly into the flue gas duct prior to a fabric filter. The limestone reacts in the hot medium with the SO2 contained in the combustion gases and is deposited on the filter bags as sodium sulfate and sodium sulfite. The dry particulate matter is then discharged to disposal and the cleaned flue gases pass through the filter medium to the stack (fig lo-lb).

System Type	SO_x Removal Efficiency (%)	Pressure Drop (inches of water)	Recovery and Regeneration	Operational Reliability	Retrofit to Existing Installation	Advantages	Disadvantages
1) Limestone, boiler injection type	30-40%	less than 6"	no recovery of limestone	High	Yes	high reliability	low efficiency, scalling in boiler and scrubber, small units only, solids disposal to landfill
2) Limestone, Scrubber injection type	30-40%	greater than 6"	no recovery of lime	High	Yes	high reliability no boiler scaling	low efficiency, scaling and plugging of nozzles and surfaces in scrubber solids disposal.
3) Lime, Scrubber injection type	90%	greater than 6"	no recovery of lime	Low	Yes	high efficiency, no boiler scaling, less scaling to scrubber than limestone in some cases	low reliability; solids disposal to landfill
4) Magnesium Oxide	90%	greater than 6"	recovery of M_g and sulfuric acid	Low	Yes	high efficiency; no solids disposal	low reliability; corrosion and erosion of scrubber and piping. Need pre-cleaning of flue gas.
5) Wellman-Lord	90%	greater than 6"	recovery of $NaSO_3$ and sulfur	Unknown	Yes	high efficiency, little scaling	unknown reliability; need natural gas for SO_2 Reduction.
6) Catalytic Oxidation	45%	may be as high as 24"	recovery of H_2SO_4	Unknown	No	high efficiency; no solids disposal problem; catalyst regeneration not necessary	Vanadium pentoxide needed as catalyst; high pressure drop system; unknown reliability; need pre-cleaning of flue gas by high efficiency device.
7) Single Alkali Systems	90%	Tray tower pressure drop 1.6-2.0 in. H_2O/tray with Venturi added 10-14 in. H_2O	little recovery of sodium	Unknown	Yes	high efficiency; reduyces scaling and plugging of scrubber	throwaway process; chemical costs high when burning high sulfur fuels; disposal of sodium salts; high water makeup rate.
8) Dual alkali Systems	90-95%		Regeneration of sodium hydroxide and sodium sulfites	Unknown	Yes	absorption efficiency potentially higher than other systems; scaling problems reduced; produces solid rather than liquid waste.	solids buildup in reactor system; problems with de-watering system.
9) Dry, post furnace limestone injection	70-80%	Greater than 6"	No recovery of limestone	High	Yes	reduced water consumption lower capital cost simplicity	most suitable for low and medium sulfer coals.
10) Dry furnace injection of LIMESTONE	40-70%	Less than 6"	No recovery of limestone	High	Yes	reduced water consumption lower capital costs simplicity	increased ash load on particulate collection equipment.

Table 10-2

Performance characteristics of flue-gas desulfurization

© J. Paul Guyer   2021

4.2.15 DRY FURNACE INJECTION OF LIMESTONE. In this system, dry ground limestone is injected into the boiler where it is calcined and reacts with the SO2 formed during combustion of the fuel. The flue gases containing the sodium sulfate, sodium sulfite, unreacted limestone, and fly ash all exit the boiler together and are captured on a particulate collector. The cleaned flue gases pass through the filter medium and out through the stack (fig 10-1c).

4.2.16 MAGNESIUM OXIDE (MGO) SCRUBBER. This is a regenerative system with recovery of the reactant and sulfuric acid. As can be seen in figure 10-2 the flue gas must be precleaned of particulate before it is sent to the scrubber. An ESP or venturi scrubber can be used to remove the particulate. The flue gas then goes to the MgO scrubber where the principal reaction takes place between the MgO and SO2 to form hydrated magnesium sulfite. Unreacted slurry is recirculated to the scrubber where it combines with makeup MgO and water and liquor from the slurry dewatering system. The reacted slurry is sent through the dewatering system where it is dried and then passed through a recovery process, the main step of which is calcination. High reliability of this system has not yet been obtained. SO2 removal efficiencies can be high, but scaling and corrosion are major problems.

4.2.17 WELLMAN LORD PROCESS. Sodium sulfite is the scrubbing solution. It captures the SO2 to produce sodium bisulfite, which is later heated to evolve SO2 and regenerate the sulfite scrubbing material. The SO2 rich product stream can be compressed or liquified and oxidized to sulfuric acid, or reduced to sulfur. Scaling and plugging are minimal problems because the sodium compounds are highly soluble in water. A Wellman- Lord unit has demonstrated an SO2 removal efficiency of greater than 90 percent and an availability of over 85 percent. The harsh acid environment of the system has caused some mechanical problems (See figure 10-3).

4.2.18 CATALYTIC OXIDATION. The catalytic oxidation process uses a high temperature (850 degrees Fahrenheit) and a catalyst (vanadium pentoxide) to convert SO2 to SO3. The heated flue gas then passes through a gas heat exchanger for heat recovery and vapor condensation. Water vapor condenses in the heat exchanger and combines with SO3 to form sulfuric acid. The acid mist is then separated from the gas in an absorbing tower. The flue gas must be precleaned by a highly efficient particulate removal device such as an electrostatic precipitator preceding the cat-ox system to avoid poisoning the catalyst. The major drawback of this system is that it cannot be economically retro-fitted to existing installations (fig 10-4).

4.2.19 SINGLE ALKALI SODIUM CARBONATE SCRUBBING. In order to eliminate the plugging and scaling problems associated with direct calcium scrubbing, this FGD system was developed. As shown in figure 10-5, the process is a once through process involving scrubbing with a solution of sodium carbonate or sodium hydroxide to produce a solution of dissolved sodium sulfur salts. The solution is then oxidized to produce a neutral solution of sodium sulfate. Because it is a throwaway process, the cost of chemicals make it an unattractive SOx removal process when burning high sulfur fuels (greater than 1 percent).

4.2.20 DUAL ALKALI SODIUM SCRUBBING.

4.2.20.1 THE DUAL ALKALI SOx removal system is a regenerative process designed for disposal of wastes in a solid/slurry form. As shown in figure 10-6, the process consists of three basic steps; gas scrubbing, a reactor system, and solids dewatering. The scrubbing system utilizes a sodium hydroxide and sodium sulfite solution. Upon absorption of SO2 in the scrubber, a solution of sodium bisulfite and sodium sulfite is produced. The scrubber effluent containing the dissolved sodium salts is reacted outside the scrubber with lime or limestone to produce a precipitate of calcium salts containing calcium sulfate. The precipitate slurry from the reactor system is dewatered and the solids are deposed of in a landfill. The liquid fraction containing soluble salts is recirculated to the absorber. Double alkali systems can achieve efficiencies of 90 - 95% and close to 100% reagent utilization.

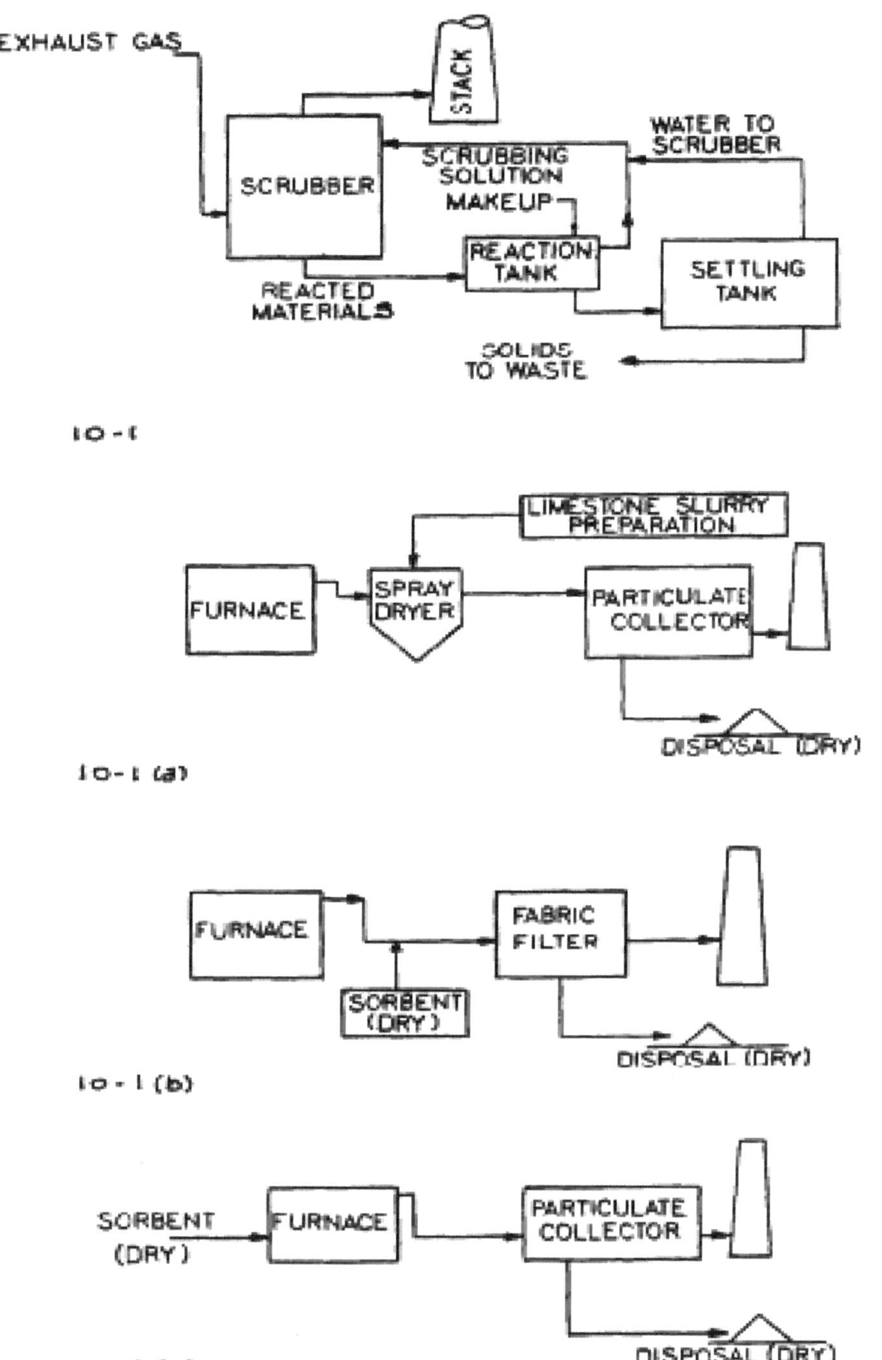

Figure 10-1

Lime (limestone) injection system schematic

© J. Paul Guyer   2021

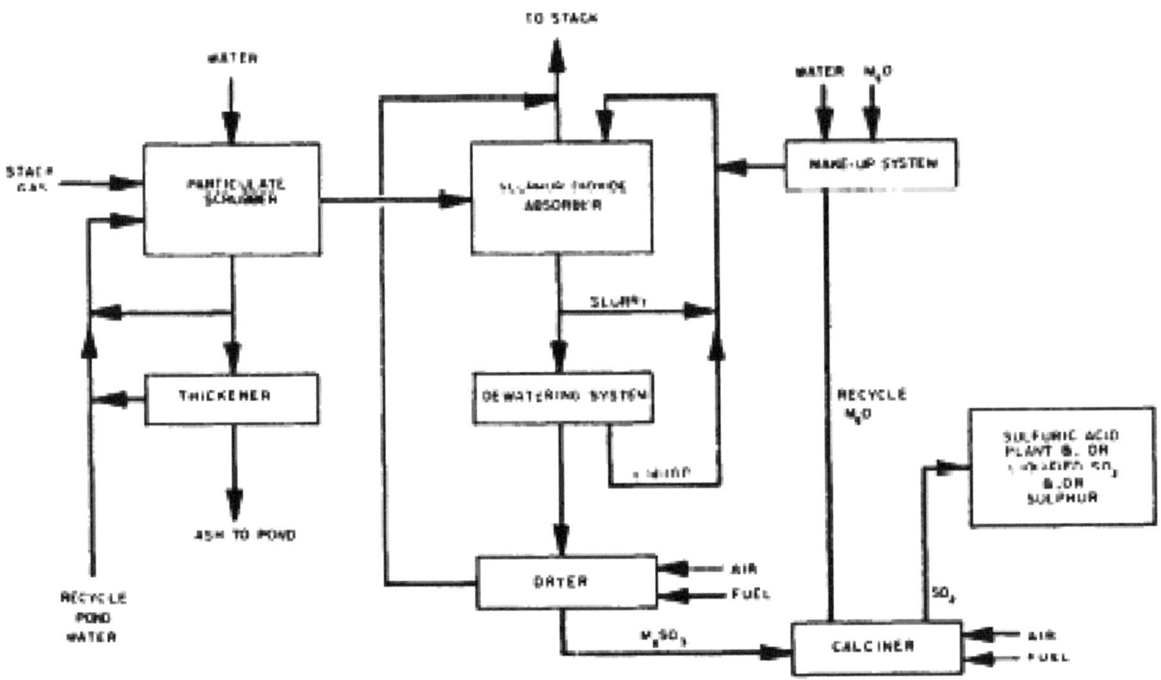

Figure 10-2

Magnesia slurry SO2 recovery process

4.2.20.2 THIS SYSTEM IS DESIGNED to overcome the inherent difficulties of direct slurry scrubbing. All precipitation occurs outside the scrubber under controlled reactor conditions. The principal advantages of the dual alkali system are:

- Scaling problems associated with direct calcium-based scrubbing processes are significantly reduced.
- A less expensive calcium base can be used.
- Due to high solubility and concentration of active chemicals, lower liquid volumes can be used thereby lowering equipment costs.
- Slurries are eliminated from the absorption loop, thereby reducing plugging and erosion problems.
- A sludge waste, rather than a liquid waste, is produced for disposal.
- High SO2 removal efficiency (90% or more).

© J. Paul Guyer   2021

4.2.21 ABSORPTION OF SO2.

4.2.21.1 ACTIVATED CARBON HAS BEEN USED as an absorbent for flue-gas desulfurization. Activated carbon affects a catalytic oxidation of SO2 to SO3, the latter having a critical temperature of 425 degrees Fahrenheit. This allows absorption to take place at operating temperatures. The carbon is subsequently regenerated in a separate reactor to yield a waste which is used in the production of high grade sulfuric acid, and the regenerated absorbent. There are serious problems involved in the regeneration of the absorbent, including carbon losses due to attrition, chemical decomposition, serious corrosion problems, and danger of combustion of the reactivated carbon.

4.2.21.2 ZEOLITES are a class of highly structured aluminum silicate compounds. Because of the regular pore size of zeolites, molecules of less than a certain critical size may be incorporated into the structure, while those greater are excluded. It is often possible to specify a certain zeolite for the separation of a particular material. Zeolites possesses properties of attrition resistance, temperature stability, inertness to regeneration techniques, and uniform pore size which make them ideal absorbents. However, they lack the ability to catalyze the oxidation of SO2 to SO3 and thus cannot desulfurize flue-gases at normal operating temperatures. Promising research is under way on the development of a zeolite material that will absorb SO2 at flue-gas temperatures by oxidation of SO3 and subsequently store it as a sulphate in the pores of the zeolite.

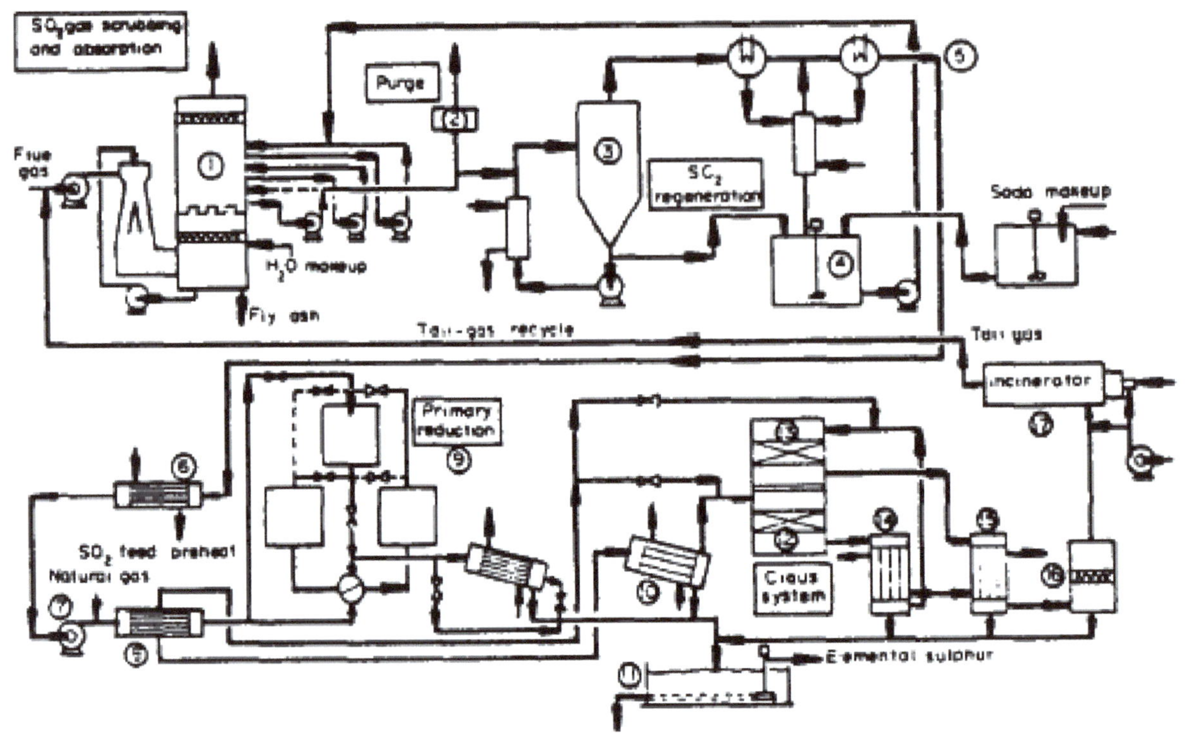

1) ABSORBER

2) SURGE TANK

3) REGENERATION SECTION

4) DISSOLVING TANK

5) CONCENTRATED SO DISCHARGE

6) PREHEATER

7) NATURAL GAS ADDITION

8) HEATER

9) CATALYTIC REDUCTION SECTION

10) COOLER

11) SULPHUR HOLDING PIT

12,13) CLAUS CONVERTER

14, 15) CONDENSORS

16) COALESCER

17) INCINERATOR

Figure 10-3

Wellman-Lord SO2 reduction system

4.2.22 COST OF FLUE-GAS DESULFURIZATION. The actual capital and operating costs for any specific installation are a function of a number of factors quite specific to the plant and include:

- -Plant size, age, configuration, and locations,

- -Sulfur content of the fuel and emission control requirements,

- -Local construction costs, plant labor costs, and cost for chemicals, water, waste disposal, etc.,

- -Type of FGD system and required equipment,

- -Whether simultaneous particulate emission reduction is required.

4.3 PROCEDURE TO MINIMIZE SOX EMISSION

4.3.1 EFFICIENCY REQUIREMENT. The SOx removal efficiency necessary for any given installation is dependent upon the strictest regulation governing that installation. Given a certain required efficiency, a choice can be made among the different reduction techniques. This section shows how a rational basis can be utilized to determine the best method.

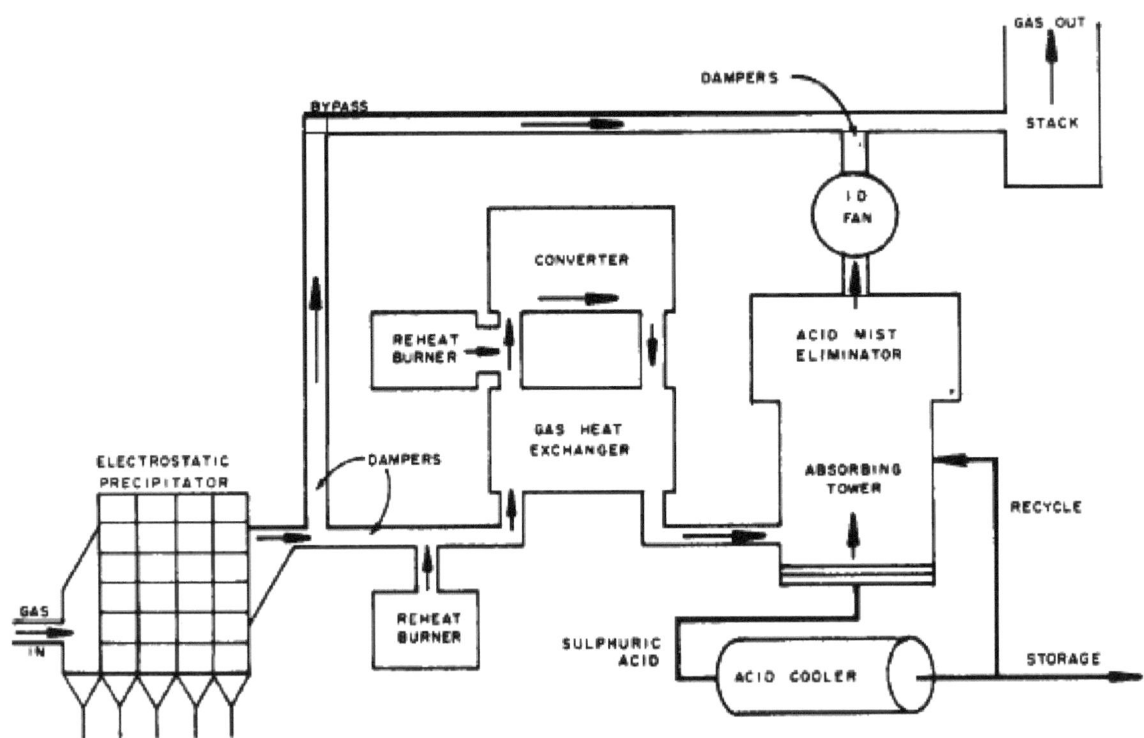

Figure 10-4

Catalytic oxidation system

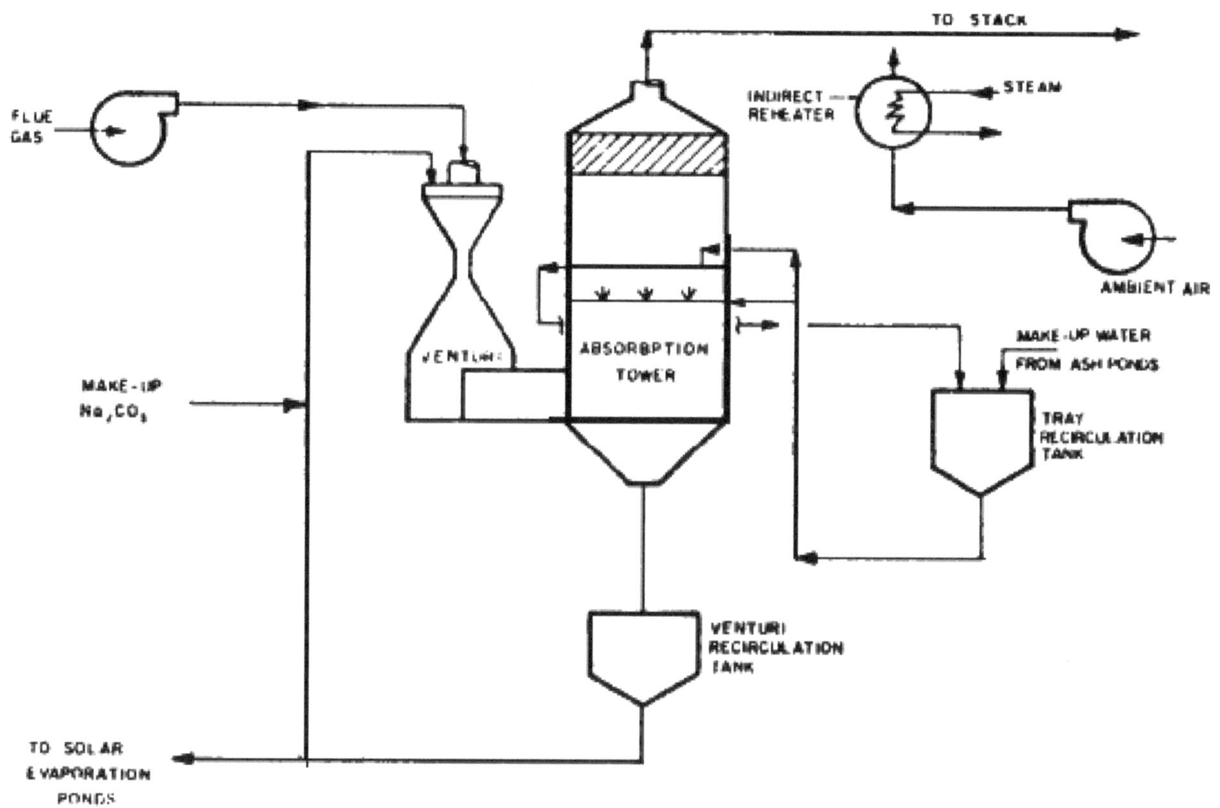

Figure 10-5

Single alkali sodium carbonate scrubber

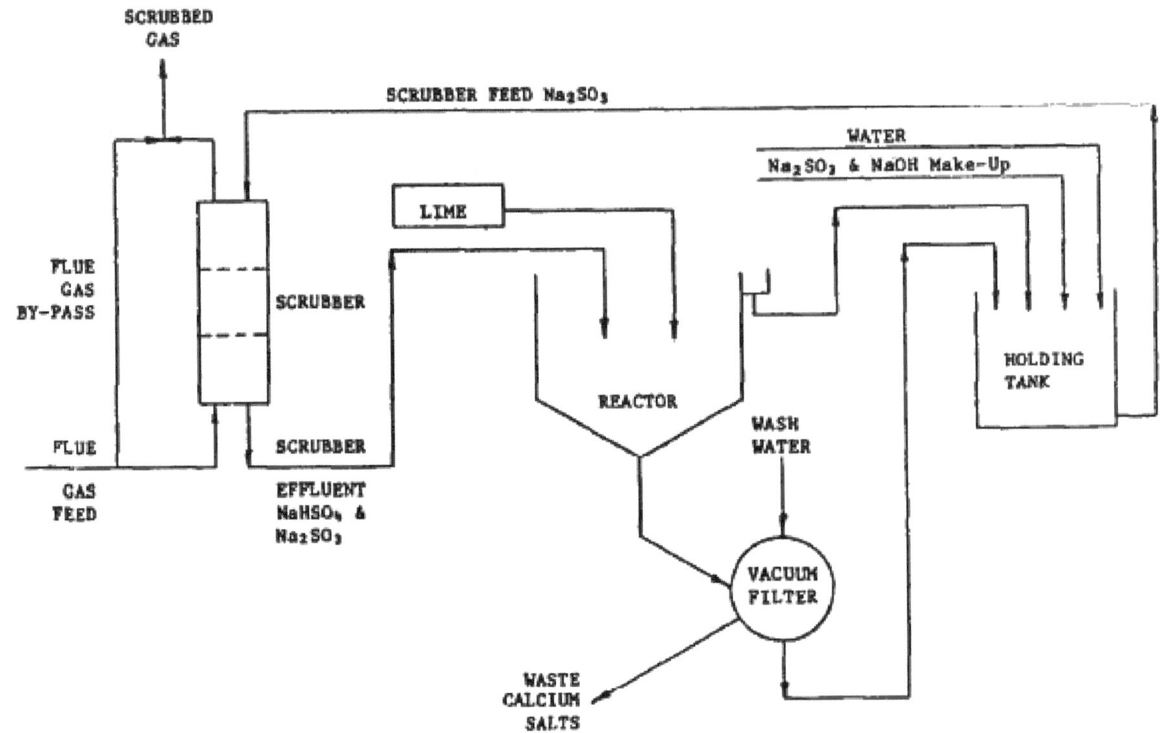

Figure 10-6

Dual alkali system

4.3.2 BOILER MODIFICATION. This technique is useful in reducing SOX emissions by 0 to 6% depending upon the boiler. For industrial boilers operating above 20% excess-air the use of proper control equipment or low excess-air combustion will usually reduce emissions by 4 to 5%. If the operating engineer is not familiar with boiler optimization methods, consultants should be utilized.

4.3.3 FUEL SUBSTITUTION. This method can be used for almost any percent reduction necessary. Availability and cost of the fuel are the major factors to be considered. Fuels can be blended to produce the desired sulfur input. Care must be taken, however, so that the ash produced by the blending does not adversely affect the boiler by lowering the ash fusion temperature or causing increased fouling in the convection banks.

4.3.4 FLUE-GAS DESULFURIZATION. Various systems are available for flue-gas desulfurization. Some of these systems have demonstrated long term reliability of operation with high SOx removal efficiency. Lime/limestone injection and scrubbing systems have been most frequently used. It must be recognized that each boiler control situation must be accommodated in the overall system design if the most appropriate system is to be installed. The selection and design of such a control system should include the following considerations:

- Local SO2 and particulate emission requirements, both present and future,
- Local liquid and solid waste disposal regulations,
- Local market demand for recovered sulfur,
- Plant design limitations and site characteristics,
- Local cost and availability of chemicals, utilities, fuels, etc.,
- Added energy costs due to process pumps, reheaters, booster fans, etc.

4.4 SAMPLE PROBLEMS. The following problems have been provided to illustrate how to determine the maximum fuel sulfur content allowable to limit SOx emission to any particular level.

4.4.1 APPROXIMATELY 90 TO 97 PERCENT OF FUEL sulfur is oxidized to sulfur dioxide (SO2) during combustion. This means that for every lb of sulfur in the fuel, approximately 2 lbs of sulfur oxides will appear in the stack gases. (The atomic weight of oxygen is l/2 that of sulfur.) Since most of the sulfur oxides are in the form of SO2, emissions regulations are defined in these units. To estimate maximum probable SO2 emissions, the following equation applies:

$$\frac{\text{lbs } SO_2}{\text{MMBtu}} = \frac{2 \times \% \text{ Sulfur by Weight in Fuel}}{\text{Fuel Heating Value (Btu/lb)}} \times 10^4$$

$$\text{(eq. 10-1)}$$

4.4.2 ASSUME A FUEL-OIL BURNING boiler must limit emissions to .35 lbs/MMBtu. What is the maximum allowable sulfur content if No. 6 Residual fuel-oil is to be used?

4.4.2.1 FROM TABLE 10-3, TYPICAL ANALYSIS OF FUEL-OIL TYPES, an average heating value of 18,300 Btu/ lb for No. 6 residual fuel has been assumed. Maximum allowable sulfur content is determined as:

$$\frac{.35 \text{ lbs } SO_2}{\text{MMBtu}}$$

$$\times \frac{2 \times \% \text{ Sulfur by Weight in Fuel}}{18,300 \text{ Btu}} \times 10^4$$

$$\% \text{ Sulfur} = \frac{(.35)(18,300)}{2 \times 10^4} = .32\%$$

4.4.2.2 TABLE 10-3 SHOWS THAT No. 5 and No. 6 fuel oils have fuel sulfur contents in excess of .32%. If No. 4 fuel oil is chosen, a fuel with less than .32% sulfur may be available.

4.4.3 ASSUME A FUEL-OIL BURNING BOILER must limit SOx emission to .65 lbs/MMBtu. If No. 6 residual fuel oil is to be used, can SOx emission limits be met?

4.4.3.1 FROM TABLE 10-3, THE MINIMUM sulfur content in No. 6 fuel oil is .7%. If .7% sulfur fuel can be purchased, the heating value of the fuel must be:

$$a) \quad .65 \ \frac{lb \ SO_2}{MMBtu} = \frac{2 \times .7\% \ Sulfur}{H. \ Value} \times 10^4$$

$$b) \ Heating \ Value = 21{,}538 \ Btu/lb$$

4.4.3.2 SINCE THE HEATING VALUE OF No. 6 fuel oil is generally between 17,410 and 18,990 Btu/lb, SOx emission limits cannot be met using this fuel. If we assume a No. 6 fuel-oil with one percent sulfur and a heating value of 18,600 Btu/lb is used the percent SOx removal efficiency that will be required is determined as:

$$a) \ \frac{lbs \ SO_2 \ emissions.}{MMBtu} \times \frac{2 \times 1\% \ Sulfur}{= 18{,}600 \ Btu/lb} \times 10^4 = 1.08$$

$$b) \ \frac{1.08 - .65}{1.08} = 40\% \ removal \ of \ SO_2 \ required$$

4.4.4 ASSUME A BOILER INSTALLATION burns No. 4 fuel-oil with a heating value of 19,000 Btu/lb. What is the maximum fuel sulfur content allowable to limit SOx emissions to .8 lbs/MMBtu?

$$(1) \quad \frac{.80 \ (\text{lbs } SO_2)}{\text{Mil Btu}} = \frac{2 \times \% \text{ Fuel Sulfur}}{19,000 \ \text{Btu/lb}} \times 10^4$$

$$(2) \ \% \text{ Fuel Sulfur} = \frac{.80 \ (19,000)}{2 \times 10^4}$$

$$= .76\%$$

4.4.5 ASSUME A COAL BURNING BOILER MUST limit SOx emissions to 1 lb/MMBtu. If sub-bituminous coal with a heating value of 12,000 to 12,500 Btu/lb (see table 10-4) is to be used what is the maximum allowable fuel sulfur content?

$$(1) \quad \frac{(1.0 \ \text{lbs } SO_2}{MMBtu} = \frac{2 \times \% \text{ Fuel Sulfur}}{12,000 \ \text{Btu/lb}} \times 10,000$$

$$(2) \ \% \text{ Fuel Sulfur} = \frac{1.0 \ (12,000)}{2 \times 10,000}$$

$$= .60\%$$

4.4.6 SINCE COAL OF THIS LOW SULFUR content is not available, what SOx removal efficiency would be required burning 1% sulfur coal?

$$(1) \quad \text{Estimated } SO_2 \text{ emissions} = \frac{\text{lbs}}{MMBtu} \ \frac{2 \times 1\%}{12,000} \times 10^4$$

$$= \frac{1.67 \ \text{lbs } SO_2}{\text{Mil Btu}}$$

$$(2) \ \% \text{ Removal Efficiency} = \frac{1.66-1.0}{1.66}$$

$$= 40\%$$

Grade of Fuel Oil	No. 1	No. 2	No. 4	No. 5	No. 6
Weight, precent					
Sulfur	0.01–0.5	0.05–1.0	0.2–2.0	0.5–3.0	0.7–3.5
Hydrogen	12.3–14.5	11.8–13.9	(10.6–13.0)	(10.5–12.0)	(86.5–90.2)
Carbon	85.9–86.7	86.1–88.2	(86.5–89.2)	(86.5–89.2)	(86.5–90.2)
Nitrogen	Nil–0.1	Nil–0.1	–	–	–
Oxygen	–	–	–	–	–
Ash	–	–	0–0.1	0–0.1	0.01–0.5
Gravity					
Deg API	40–44	28–40	15–30	14–22	7–22
Specific	0.825–0.806	0.887–0.825	0.996–0.876	0.972–0.922	1.022–0.922
Lb per gal	6.87–6.71	7.39–6.87	8.04–7.30	8.10–7.68)	8.51–7.68
Pour point, F	0 to –50	0 to –40	–10 to +50	–10 to +80	+15 to +85
Viscosity					
Centistokes at 100°F	1.4–2.2	1.9–3.0	10.5–65	65–200	260–750
SUS at 100°F	–	32–38	60–300	–	–
SSF at 122°F	–	–	–	20–40	45–300
Water and Sediment, Vol%	–	0–0.1	tr–1.0	0.05–1.0	0.05–2.0
Heating Value					
Btu, per lb, gross	19,670–19,860	19,170–19,750	18,280–19,400	18,100–19,020	17,410–18,990

Table 10-3

Typical analysis of Fuel Oil Types

Source	Anthracite	Bituminous	Subbitiminous	Lignite
	Pennsylvania Arkansas	Pennsylvania West Virginia	Wyoming	Texas North Dakota
Moisture (as received	4-6%	4-12%	12-20%	20-40%
Proximate Analysis (Dry Basis)				
Volatile Matter	3-12%	16-40%	40-45%	45-50%
Fixed Carbon	75-84%	50-80%	45-55%	35-45%
Ash (average)	14-15%	4- 9%	5-14%	10-12%
Sulfur	1- 2%	1- 2%	1- 4%	1- 2%
Ultimate Analysis (Dry Basis)				
Hydrogen	.5-3.5%	5%	4- 5%	4-4.5%
Carbon	75-84%	73-85%	60-73%	65%
Nitrogen	.1-.5%	1.5%	.9-1.3%	1.2-1.9%
Oxygen	1- 3%	3-13%	14-16%	15-18%
Heating Value	11,500-13.000	13,000-15,000	10,500-13,000	11,000-12,000
Aggolomerating Character	non- agglomerating	Commonly agglomerating	non- agglomerating	non- agglomerating
Weathering Character	-	-	-	weathering
Relative Hardness	Hard	Granular	Sandlike	Soft and Fibrous
Method of Firing				
Traveling Grate Stoker	X	-	X	X
Spreader Stokers	-	X	X	X
Pulverized Coal Burners	X	X	X	X

Table 10-4

Typical analysis of coal types

4.5 NITROGEN OXIDES (NOx) CONTROL AND REDUCTION, TECHNIQUES

4.5.1 FORMATION OF NITROGEN OXIDES

4.5.1.1 NITROGEN OXIDES (NOx). All fossil fuel burning processes produce NOx. The principle oxides formed are nitric oxide (NO) which represents 90-95 percent (%) of the NOx formed and nitrogen dioxide (NOx) which represents most of the remaining nitrogen oxides.

4.5.1.2 NOx FORMATION Nitrogen oxides are formed primarily in the high temperature zone of a furnace where sufficient concentrations of nitrogen and oxygen are present. Fuel nitrogen and nitrogen contained in the combustion air both play a role in the formation of NOx. The largest percentage of NOx formed is a result of the high temperature fixation reaction of atmospheric nitrogen and oxygen in the primary combustion zone.

4.5.1.3 NOx CONCENTRATION. The concentration of NOx found in stack gas is dependent upon the time, temperature, and concentration history of the combustion gas as it moves through the furnace. NOx concentration will increase with temperature, the availability of oxygen, and the time the oxygen and nitrogen simultaneously are exposed to peak flame temperatures.

4.5.2 FACTORS AFFECTING NOx EMISSIONS

4.5.2.1 FURNACE DESIGN AND FIRING TYPE. The size and design of boiler furnaces have a major effect on NOx emissions. As furnace size and heat release rates increase, NOx emissions increase. This results from a lower furnace surface-to-volume ratio which leads to a higher furnace temperature and less rapid terminal quenching of the combustion process. Boilers generate different amounts of NOx according to the type of firing. Units employing less rapid and intense burning from incomplete mixing of fuel and combustion gases generate lower levels of NOx emissions. Tangentially tied units

© J. Paul Guyer   2021

generate the least NOx because they operate on low levels of excess air, and because bulk misting and burning of the fuel takes place in a large portion of the furnace. Since the entire furnace acts as a burner, precise proportioning of fuel/air at each of the individual fuel admission points is not required. A large amount of internal recirculation of bulk gas, coupled with slower mixing of fuel and air, provides a combustion system which is inherently low in NOx production for all fuel types.

4.5.2.2 BURNER DESIGN AND CONFIGURATION. Burners operating under highly turbulent and intense flame conditions produce more NOx. The more bulk mixing of fuel and air in the primary combustion zone, the more turbulence is created. Flame color is an index of flame turbulence. Yellow hazy flames have low turbulence, whereas, blue flames with good definition are considered highly turbulent.

4.5.2.3 BURNER NUMBER. The number of burners and their spacing are important in NOx emission. Interaction between closely spaced burners especially in the center of a multiple burner installation, increases flame temperature at these locations. The tighter spacing lowers the ability to radiate to cooling surfaces, and greater is the tendency toward increased NOx emissions.

4.5.2.4 EXCESS AIR. A level of excess air greatly exceeding the theoretical excess air requirement is the major cause of high NOx emissions in conventional boilers. Negotiable quantities of thermally formed NOx are generated in fluidized bed boilers.

4.5.2.5 COMBUSTION TEMPERATURE. NOx formation is dependent upon peak combustion temperature, with higher temperatures producing higher NOX emissions.

4.5.2.6 FIRING AND QUENCHING RATES. A high heat release rate (firing rate) is associated with higher peak temperatures and increased NOx emissions. A high rate of thermal quenching, (the efficient removal of the heat released in combustion) tends to lower peak temperatures and contribute to reduced NOx emissions.

4.5.2.7 MASS TRANSPORTATION AND MIXING. The concentration of nitrogen and oxygen in the combustion zone affects NOx formation. Any means of decreasing the concentration such as dilution by exhaust gases, slow diffusion of fuel and air, or alternate fuel-rich/fuel-lean burner operation will reduce NOx formation. These methods are also effective in reducing peak flame temperatures.

4.5.2.8 FUEL TYPE. Fuel type affects NOx formation both through the theoretical flame temperature reached, and through the rate of radiative heat transfer. For most combustion installations, coal-fired furnaces have the highest level of NOx emissions and gas-fired installations have the lowest levels of NOx emissions.

4.5.2.9 FUEL NITROGEN. The importance of chemically bound fuel nitrogen in NOx formation varies with the temperature level of the combustion processes. Fuel nitrogen is important at low temperature combustion, but its contribution is nearly negligible as higher flame temperatures are reached, because atmospheric nitrogen contributes more to NOx formation at higher temperatures.

4.5.3 NOx REDUCTION TECHNIQUES

4.5.3.1 FUEL SELECTION. Reduction of NOx emissions may be accomplished by changing to a fuel which decreases the combustion excess air requirements, peak flame temperatures, and nitrogen content of the fuel. These changes decrease the concentration of oxygen and nitrogen in the flame envelope and the rate of the NOx formation reaction.

4.5.3.1.1 THE SPECIFIC BOILER MANUFACTURER should be consulted to determine if a fuel conversion can be performed without adverse effects. The general NOx reduction capability of initiating a change in fuel can be seen comparatively in table 11-l.

4.5.3.1.2 A CONSIDERATION WHEN contemplating a change in fuel type is that NOx emission regulations are usually based on fuel type. Switching to a cleaner fuel may result in the necessity of conforming to a more strict emission standard.

4.5.3.1.3 CHANGING FROM A higher to a lower NOx producing fuel is not usually an economical method of reducing NOx emissions because additional fuel costs and equipment capital costs will result.

Fuel Type	Range of Excess Air Level, percent above stoichiometric	NO_x Emissions lbs/Mil Btu
coal	18-25	.5-1.1
(Lignite)	-	(.9-1.1)
Fuel Oil	3-15	.1- .4
Gaseous fuels	7-10	.3

Table 11-1

General NOx emission and excess air requirements for fuel types

4.5.3.2 LOAD REDUCTION. Load reduction is an effective technique for reducing NOx emissions. Load reduction has the effect of decreasing the heat release rate and reducing furnace temperature. A lowering of furnace temperature decreases the rate of NOx formation.

4.5.3.2.1 NOX REDUCTION BY LOAD REDUCTION is illustrated in figure 11-l. As shown, a greater reduction in NO2 is attainable burning gas fuels because they contain only a small amount of fuel-bound nitrogen. Fuel-bound nitrogen conversion does not appear to be affected by furnace temperatures, which accounts for the lower NOx reductions obtained with coal and oil tiring. Some units such as tangentially fired boilers show as much as 25 percent decrease in NOx emissions with a 25 percent load reduction while burning pulverized coal.

© J. Paul Guyer   2021

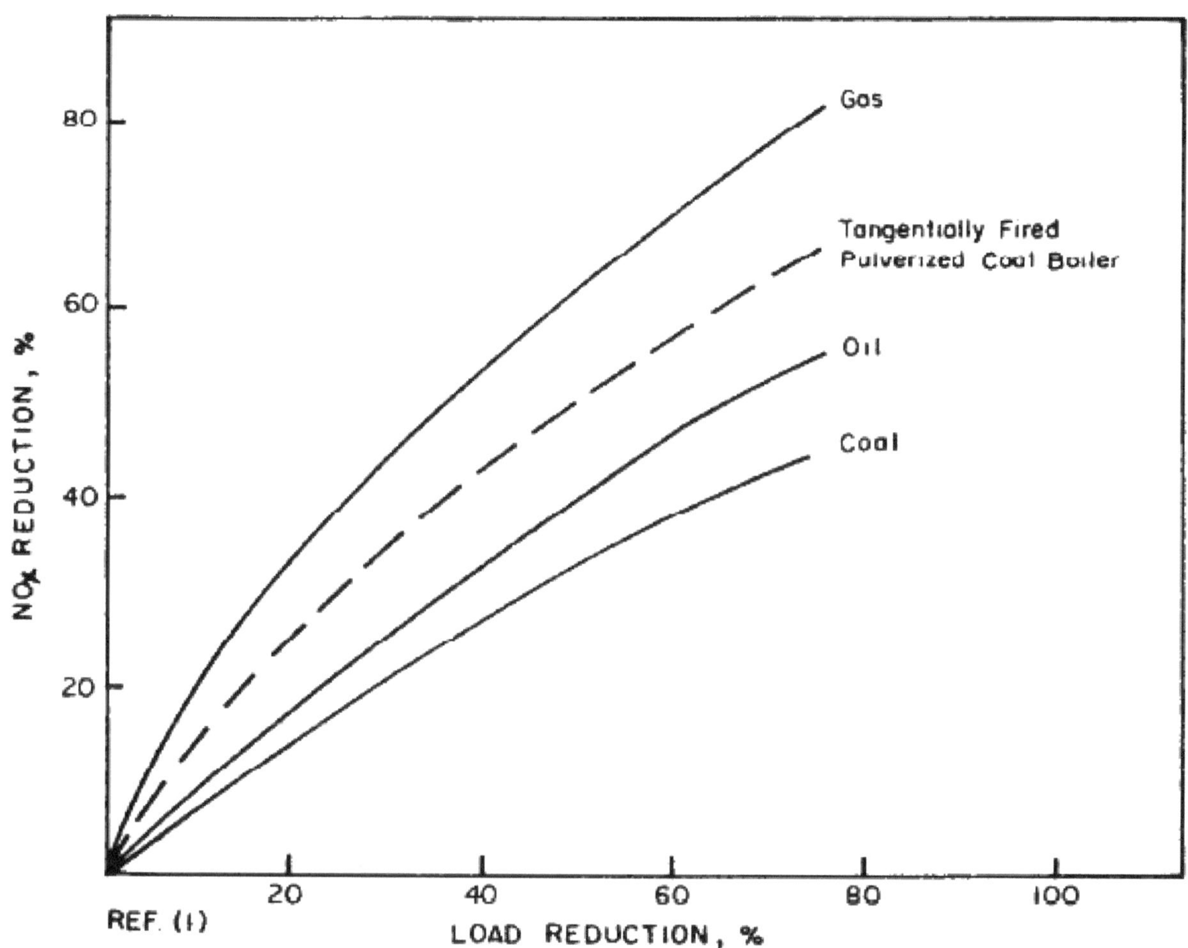

Figure 11-1

Possible NOx reductions vs load reductions

4.5.3.2.2 ALTHOUGH NO CAPITAL COSTS are involved in load reduction, it is sometimes undesirable to reduce load because it may reduce steam cycle efficiency.

4.5.3.3 LOW EXCESS AIR FIRING (LEA). In order to complete the combustion of a fuel, a certain amount of excess air is necessary beyond the stoichiometric requirements. The more efficient the burners are in misting, the smaller will be the excess air requirement. A minimum amount of excess air is needed in any system to limit the production of smoke or unburned combustibles; but larger amounts may be needed to maintain steam temperature to prevent refractory damage; to complete

© J. Paul Guyer 2021

combustion when air supply between burners is unbalanced; and to compensate for instrument lag between operational changes. Practical minimums of excess air are 7 percent for natural gas, 3 to 15 percent for oil firing, and 18 to 25 percent for coal firing.

4.5.3.3.1 SINCE AN INCREASE IN the amount of oxygen and nitrogen in a combustion process will increase the formation and concentration of NOx, low excess air operation is the first and most important technique that should be utilized to reduce NOx emissions. A 50 percent reduction in excess air can usually reduce NOx emissions from 15 to 40 percent, depending upon the level of excess air normally applied. Average NOx reductions corresponding to a 50 percent reduction in excess air for each of the three fuels in different boiler types are shown in table 11-2. Reductions in NOx emissions up to 62 percent have been reported on a pulverized coal fired boiler when excess air is decreased from a level of 22 percent to a level of 5 percent.

| | Fuel Type | | |
| | Gas | Oil | Coal |
Furnace Type	% Reduction		
Horizontal front wall	15-20	20-25	25-35
Horizontal opposed firing	15-25	25-30	30-40
Tangentially fired	15	20-25	25-30

NOTE: Overall NO_x reduction potential for industrial and commercial size boilers for LEA operation is limited to about 40% reduction dependent upon the level of excess air normally applied.

Table 11-2

Possible NOx emission reductions attainable with a 50% reduction in excess air from normal levels (greater than 10% excess air)

4.5.3.3.2 THE SUCCESSFUL APPLICATION of LEA firing to any unit requires a combustion control system to regulate and monitor the exact proportioning of fuel and air. For pulverized coal fired boilers, this may mean the additional expense of installing uniform distribution systems for the coal and air mixture.

4.5.3.3.3 LOW EXCESS AIR FIRING IS a desirable method of reducing NOx emission because it can also improve boiler efficiency by reducing the amount of heat lost up the stack. Consequently, a reduction in fuel combustion will sometimes accompany LEA firing.

4.5.3.4 LOW EXCESS AIR FIRING WITH LOAD REDUCTION. NOx emissions may be reduced by implementing a load reduction while operating under low excess air conditions (table 11-2). This combined technique may be desirable in an installation where NOx emissions are extremely high because of poor air distribution and the resultant inefficient operation of combustible equipment. A load reduction may permit more accurate control of the combustion equipment and allow reduction of excess air requirements to a minimum value. NOx reduction achieved by simultaneous implementation of load reduction and LEA firing is slightly less than the combined estimated NOx reduction achieved by separate implementation.

4.5.3.5 TWO-STAGE COMBUSTION. The application of delayed fuel and air mixing in combustion boilers is referred to as two stage combustion. Two-stage combustion can be of two forms. Normally it entails operating burners fuel-rich (supplying only 90 to 95 percent of stoichiometric combustion air) at the burner throat, and admitting the additional air needed to complete combustion through ports (referred to as NO ports) located above and below the burner. There are no ports to direct streams of combustion air into the burner flame further out from the burner wall thus allowing a gradual burning of all fuel. Another form of two-stage combustion is off-stoichiometric firing. This technique involves firing some burners fuel-rich and others airrich (high percentage of excess air), or air only, and is usually applied to boilers having three or more burner levels. Off-stoichiometric firing is accomplished by staggering the air-rich and fuel-rich burners in each of the burner levels. Various burner configuration tests have shown that it is generally more effective to operate most of the elevated burners air-rich or air only.

Off-stoichiometric firing in pulverized coal fired boilers usually consists of using the upper burners on air only while operating the lower levels of burners fuel-rich. This technique is called over-fire air operation.

4.5.3.5.1 TWO-STAGE COMBUSTION IS EFFECTIVE in reducing NOx emissions because: it lowers the concentration of oxygen and nitrogen in the primary combustion zone by fuel-rich firing; it lowers the attainable peak flame temperature by allowing for gradual combustion of all the fuel; and it reduces the amount of time the fuel and air mixture is exposed to higher temperatures.

4.5.3.5.2 THE APPLICATION OF SOME FORM OF two stage combustion implemented with overall low excess air operation is presently the most effective method of reducing NOx emissions in utility boilers. Average NOx reductions for this combustion modification technique in utility boilers are listed in table 11-3. However, it should be noted that this technique is not usually adaptable to small industrial boilers where only one level of burners is provided.

	Fuel Type		
	Gas	Oil	Coal
Furnace Type	Percent Reduction		
Two-stage combustion			
Range	40–70	20–50	20–40
Average	50	40	35
Off-stoichiometric	–	–	39–60
Two-stage combustion or off-stoichiometric with LEA or load reduction			
Range	50–90	40–70	40–60
Average	70	60	50

Table 11-3

Possible NOx reductions, percent of normal emissions

© J. Paul Guyer   2021

4.5.3.6 REDUCED PREHEAT TEMPERATURE. NOx emissions are influenced by the effective peak temperature of the combustion process. Any modifications that lower peak temperature will lower NOx emissions. Lower air preheat temperature has been demonstrated to be a factor in controlling NOx emissions. However, reduced preheat temperature is not a practical approach to NOx reduction because air preheat can only be varied in a narrow range without upsetting the thermal balance of the boiler. Elimination of air preheat might be expected to increase particulate emissions when burning coal or oil. Preheated air is also a necessary part of the coal pulverizer operation on coal fired units. In view of he penalties of reduced boiler efficiency and other disadvantages, reduced preheat is not a preferred means of lowering NOx emissions.

4.5.3.7 FLUE-GAS RECIRCULATION. This technique is used to lower primary combustion temperature by recirculating part of the exhaust gases back into the boiler combustion air manifold. This dilution not only decreases peak combustion flame temperatures but also decreases the concentration of oxygen available for NOx formation. NOx reductions of 20 to 50 percent have been obtained on oil-fired utility boilers but as yet have not been demonstrated on coal-fired units. It is estimated that flue gas recirculation has a potential of decreasing NOx emissions by 40 percent in coal-fired units.

4.5.3.7.1 FLUE GAS RECIRCULATION has also produced a reduction on CO concentrations from normal operation because of increased fuel-air mixing accompanying the increased combustion air/ gas volume. Gas recirculation does not significantly reduce plant thermal efficiency but it can influence boiler operation. Radiation heat transfer is reduced in the furnace because of lower gas temperatures, and convective heat transfer is increased because of greater gas flow.

4.5.3.7.2 THE EXTENT OF THE applicability of this modification remains to be investigated. The quantity of gas necessary to achieve the desired effect in different installations is important and can influence the feasibility of the application.

© J. Paul Guyer   2021

Implementing flue-gas recirculation means providing duct work and recycle fans for diverting a portion of the exhaust flue-gas back to the combustion air windbox. It also requires enlarging the windbox and adding control dampers and instrumentation to automatically vary flue-gas recirculation as required for operating conditions and loads.

4.5.3.8 STEAM OR WATER INJECTION. Steam and water injection has been used to decrease flame temperatures and reduce NOx emissions. Water injection is preferred over steam because of its greater ability to reduce temperature. In gas and coal fired units equipped with standby oil firing with steam atomization, the atomizer offers a simple means for injection. Other installations require special equipment and a study to determine the proper point and degree of atomization. The use of water or steam injection may entail some undesirable operating conditions, such as decreased efficiency and increased corrosion: A NOx reduction rate of up to 10 percent is possible before boiler efficiency is reduced to uneconomic levels. If the use of water injection requires installation of an injection pump and attendant piping, it is usually not a cost-effective means of reducing NOx emissions.

4.5.4 POST COMBUSTION SYSTEMS FOR NOx REDUCTION.

4.5.4.1 SELECTIVE CATALYTIC REDUCTION (SCR) of NOx is based on the preference of ammonia to react with NO, rather than with other flue-gas constitutents. Ammonia is injected so that it will mix with flue-gas between the economizer and the air heater. Reaction then occurs as this mix passes through a catalyst bed. Problems requiring resolution include impact of ammonia on downstream equipment, catalyst life, fluegas monitoring, ammonia availability, and spent-catalyst disposal.

4.5.4.2 SELECTIVE NONCATALYTIC REDUCTION (SNR) Ammonia is injected into the flue-gas duct where the temperature favors the reaction of ammonia with NOx in the flue gas. The narrow temperature band which favors the reaction and the difficulty of controlling the temperature are the main drawbacks of this method. c. Copper oxide is used as the acceptor for SO, removal, forming copper sulfate. Subsequently both the copper sulfate which was formed and the copper oxide catalyze the reduction of NO to nitrogen and water by reaction with ammonia. A regeneration step produces an SO, rich steam which can be used to manufacture by-products such as sulfuric acid.

4.5.5 STEP-BY-STEP NOX REDUCTION METHOD

4.5.5.1 APPLICABILITY. The application of NOx reduction techniques in stationary combustion boilers is not extensive. (However, NOx reduction techniques have been extensively applied on automobiles.) These techniques have been confined to large industrial and utility boilers where they can be more easily implemented where NOx emissions standards apply, and where equipment modifications are more economically justified. However, some form of NOx control is available for all fuel-burning boilers without sacrificing unit output or operating efficiency. Such controls may become more widespread as emission regulations are broadened to include all fuel-burning boilers.

4.5.5.2 IMPLEMENTATION. The ability to implement a particular combustion modification technique is dependent upon furnace design, size, and the degree of equipment operational control. In many cases, the cost of conversion to implement a modification such as flue-gas recirculation may not be economically justified. Therefore, the practical and economic aspects of boiler design and operational modifications must be ascertained before implementing a specific reduction technique.

4.5.5.2.1 TEMPERATURE REDUCTION THROUGH the use of two-stage combustion and flue-gas recirculation is most applicable to high heat release boilers with a multiplicity of burners such as utility and large industrial boilers.

4.5.5.2.2 LOW EXCESS AIR OPERATION (LEA) coupled with flue-gas recirculation offers the most viable solution in smaller industrial and commercial size boilers. These units are normally designed for lower heat rates (furnace temperature) and generally operate on high levels of excess air (30 to 60%).

4.5.5.3 COMPLIANCE. When it has been ascertained that NOx emissions must be reduced in order to comply with state and federal codes, a specific program should be designed to achieve the results desired. The program direction should include: -an estimate of the NOx reduction desired, -selection of the technique or combination thereof, which will achieve this reduction; -an economic evaluation of implementing each technique, including equipment costs, and changes in operational costs; -required design changes to equipment -the effects of each technique upon boiler performance and operational safety.

4.5.5.4 PROCEDURE. A technical program for implementing a NOx reduction program should proceed with the aid of equipment manufacturers and personnel who have had experience in implementing each of the NOx reduction techniques that may be required in the following manner:

4.5.5.4.1 NOx EMISSION TEST. A NOx emission test should be performed during normal boiler load times to ascertain actual on-site NOx generation. This test should include recording of normal boiler parameters such as: flame temperature; excess air; boiler loads; flue-gas temperatures; and firing rate. These parameters can be referred to as normal operating parameters during subsequent changes in operation.

4.5.5.4.2 REDUCTION CAPABILITIES. The desired reduction in NOx emissions, in order to comply with standards, should be estimated based on measured NOx emission data. Specific NOx reduction techniques can then be selected based on desired reductions and reduction capabilities outlined in preceding paragraph 11-3.

4.5.5.4.3 EQUIPMENT OPTIMIZATION. Any realistic program for NOx reduction should begin with an evaluation and overhaul of all combustion related equipment. A general improvement of boiler thermal efficiency and combustion efficiency will reduce the normal level of NOx emissions. Of major importance are:

- the cleanliness of all heat transfer surfaces (especially those exposed to radiative heat absorption),

- maintaining proper fuel preparation (sizing, temperature, viscosity),

- insuring control and proper operation of combustion equipment (burners nozzles, air registers, fans, preheaters, etc.),

- maintaining equal distribution of fuel and air to all burners.

4.5.5.4.4 LOW EXCESS AIR OPERATION. Low excess air operation is the most recommended modification for reducing NOx emission. Possible reductions are given in preceding table 11-2. However, a control system is needed to accurately monitor and correct air and fuel flow in response to steam demands. Of the control systems available, a system incorporating fuel and air metering with stack gas O2 correction will provide the most accurate control. A system of this nature will generally pay for itself in fuel savings over a 2 to 3-year period, and is economically justified on industrial boilers rated as low as 40,000 lb of steam/hr.

4.5.5.4.5 FLUE-GAS RECIRCULATION. Flue-gas recirculation is the second most effective NOx reduction technique for boilers where two stage combustion cannot be applied. Low excess air operation and flue-gas recirculation must be implemented simultaneously from a design point of view. LEA operation may require installation or retrofitting of air registers to maintain proper combustion air speed and mixing at reduced levels or air flow. Flue gas recirculation will require larger air registers to accommodate the increased volume of flow. Therefore, simultaneous application of LEA operation and flue-gas recirculation may minimize the need for redesign of burner air registers. Knowledge of furnace thermal design must accompany any application of flue-gas recirculation which effectively lowers furnace temperature and thus, radiative heat transfer. Convective heat transfer is also increased by increased gas flow due to the dilution of combustion air. It is advisable to consult boiler manufacturers as to the applicability of flue-gas recirculation to their furnaces.

© J. Paul Guyer   2021

CHAPTER 5
AIR STRIPPING

5.1 INTRODUCTION

5.1.1 PURPOSE. This discussion provides technical information for designing and constructing air stripping systems.

5.1.2 DEFINITION. Air strippers remove volatile organic chemicals (VOCs) from liquid (water) by providing contact between the liquid and gas (air). The gas (air) may then be released to the atmosphere or treated to remove the VOCs and subsequently released to the atmosphere.

5.1.3 SCOPE. This discussion describes packed column, low-profile sieve tray and diffused aeration air strippers. Steam stripping is not included. This document discusses the three types, compares them, and lists advantages and disadvantages of each type to provide information for selection. Design examples for the packed-column and the low-profile air stripper are included in the appendices.

5.1.4 THEORY. Air stripping is the mass transfer of VOCs that are dissolved in water from the water phase to the air phase. The equilibrium relationship is linear and is defined by Henry's Law. For low concentrations of volatile compound a:

$$p_a = H_a \times X_a$$

5.1.4.1 AT EQUILIBRIUM, THE PARTIAL pressure of a gas, p_a, above a liquid is directly proportional to the mole fraction of the gas, x_a, dissolved in the liquid. The proportionality constant, H_a, is known as the Henry's constant. The value of the constant generally increases or decreases with the liquid temperature. As a consequence, the solubility of gases generally decreases with increasing temperature.

© J. Paul Guyer   2021　　　　127

EPA (1998) has published a comprehensive document, Henry's Law Constants, F_m Values, F_r Values and F_e Values for Organic Compounds, at:

- http://www.epa.gov/ttn/oarpg/tl/fr_notices/appj.pdf

Practical application of the technology for contaminant removal is generally limited to compounds with Henry's constant values greater than 100 atmospheres.

Contaminant	[atm-m³/mole]	[Pa-m³/mole]	[Dimensionless]
2,4 - D	1.02×10^{-8}	1.03×10^{-3}	7.65×10^{-9}
alachlor	3.20×10^{-8} to 1.20×10^{-10}	3.24×10^{-3} to 1.22×10^{-5}	2.40×10^{-8} to 9.00×10^{-11}
aldicarb, aldicarb sulfone and aldicarb sulfoxide	1.50×10^{-10}	1.52×10^{-4}	1.13×10^{-9}
atrazine	2.63×10^{-9}	2.66×10^{-4}	1.97×10^{-9}
carbofuran	1.02×10^{-10}	1.03×10^{-5}	7.65×10^{-11}
chlordane (gamma-chlordane)	1.30×10^{-3}	132	9.75×10^{-4}
dalapon	6.30×10^{-8}	6.38×10^{-3}	4.73×10^{-8}
dibromochloropropane (DBCP)	1.47×10^{-4}	14.9	1.10×10^{-4}
di (2-ethylhexyl) adipate	4.34×10^{-7}	4.40×10^{-2}	3.26×10^{-7}
di (2-ethlhexyl) phthalate (DEHP)	1.00×10^{-4}	10.1	7.50×10^{-5}
dinoseb	5.04×10^{-4}	51.1	3.78×10^{-4}
dioxin (2,3,7,8-TCDD)	1.62×10^{-5}	1.64	1.22×10^{-5}
endrin	4.00×10^{-7}	4.05×10^{-2}	3.00×10^{-7}
hexachlorobenzene (HCB)	3.00×10^{-2} to 7.00×10^{-2}	$3.04\times10^{+3}$ to $7.09\times10^{+3}$	2.25×10^{-2} to 5.25×10^{-2}
heptachlor and heptachlor epoxide	2.62×10^{-3}	265	1.97×10^{-3}
hexachlorocyclopentadiene (hex)	2.70×10^{-2}	$2.74\times10^{+3}$	2.03×10^{-2}
methoxychlor	1.60×10^{-5}	162	1.20×10^{-5}
polychlorinated biphenyls (PCBs)	5.00×10^{-5} to 3.30×10^{-4}	5.1 to 33.44	3.75×10^{-5} to 2.48×10^{-4}
simazine	4.63×10^{-10}	4.69×10^{-5}	3.47×10^{-10}
toxaphene	5.00×10^{-3} to 6.30×10^{-2}	507 to $6.38\times10^{+3}$	3.75×10^{-3} to 4.73×10^{-2}
http://www.epa.gov/OGWDW/dwh/t-soc/chemical name (i.e. 24-D, alachlor…).html			
The molar density of water at 293.160 K, $C_o = 55.41$ kg-mole/m³			
The universal gas constant, $R = 8.3145$ Pa-m³/kg mol-K			
$1\ Pa = 9.86923 \times 10^{-6}$ atm			

Table 1-1

Molar Henry's Law Constants at 293.16 K

5.1.4.2 UNITS, AS DEFINED BY HENRY'S LAW, as stated, are standard atmospheres [atm] with the concentration n of the solute given as the mole fraction of the solution. Practical application of Henry's law has resulted in corruption of the units to the point of confusion, as seen in Table 1-1.

© J. Paul Guyer   2021

5.2 DESCRIPTION OF AIR STRIPPERS

5.2.1 PACKED COLUMN AIR STRIPPERS. Air strippers provide contact between air and water that encourages volatile materials to move from the water to the air. A packed column air stripper consists of a cylindrical column that contains a water distribution system above engineered (structured or dumped) packing with an air distributor below (see Figure 2-1). Water containing VOCs is distributed at the top of the column and flows generally downward through the packing material. At the same time, air, introduced at the bottom of the column, flows upward through the packing (countercurrent flow). The packing provides an extended surface area and impedes the flow of both fluids, extending the contact between them. As water and air contact, VOCs move from the water to the air. The water leaves the bottom of the column depleted of VOCs. The VOCs transferred to the air exit the top of the column in the air stream. Off-gas (air) is released to the atmosphere re or treated if necessary to meet emission limits.

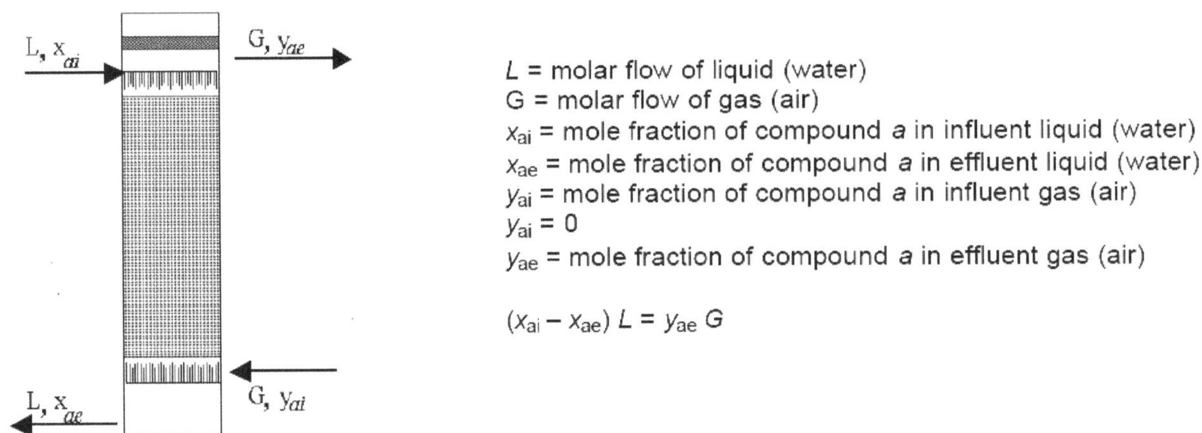

Figure 2-1

Packed column air stripper.

5.2.2 SIEVE TRAY AIR STRIPPER. Sieve tray air strippers operate in a similar way to packed column air strippers (Figure 2-2). The difference is that the liquid (water) flows across trays that are perforated with small holes, over a weir, and through a downcomer, to the next lower tray, tray by tray, until the treated water flows from the bottom of the stripper. Gas (air) is bubbled through the holes in the trays, stopping the liquid from dripping through them. The VOCs are transferred from the liquid to the gas phase as the air is bubbled through the water on the trays. Detailed information on sieve tray units is available in the literature.

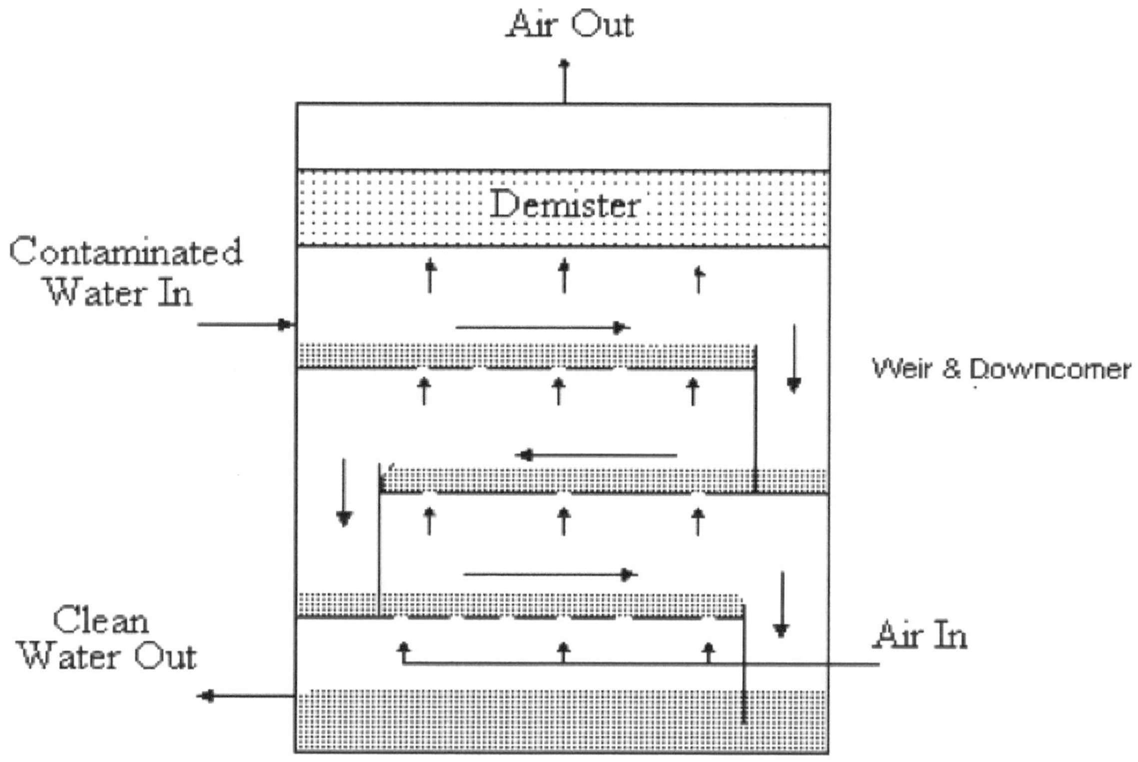

Figure 2-2

Low profile sieve tray air stripper.

5.2.3 DIFFUSED AERATION STRIPPER. A diffused aeration stripper is a vessel or liquid (water) reservoir with gas (air) diffusers near the bottom (Figure 2-3). Air enters through diffusers and rises through the liquid to exit at the top of the vessel. The VOCs move from the water to the air as the bubbles rise through the water. Transfer of the VOCs from the water to the air can be improved by increasing the vessel depth or by producing smaller bubbles. The air path through the liquid is straight and contact between the air and water is short. Therefore, diffused air is not efficient. Its main advantages are that it is simple and that it can handle water having high levels of suspended solids. Information on diffused aeration is available in the literature.

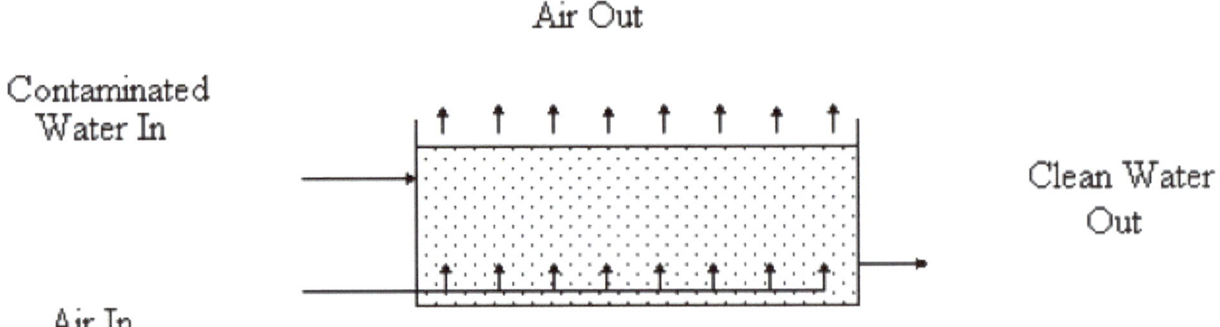

Figure 2-3

Diffused aeration air stripper.

5.3 DESIGN METHODS

5.3.1 GENERAL. The first step in designing an air stripper is to determine the extreme operating conditions: VOC concentrations allowed in the effluent, VOC concentrations in the influent, minimum liquid temperature, and influent flow rate (minimum and maximum). The next step is to calculate the total daily contaminant loading to the stripper and verify if the loading exceeds any regulation for discharge in the off- gas. A cost comparison (including water pre-treatment and off- gas treatment) is used to determine the optimum type of air stripper (packed column, low profile, or diffused aeration). Zoning regulations may limit stripper or stack height, or both. Manufacturers' software, commercial software, analytical equations, or graphical methods are used to size the air stripper, and are listed below.

5.3.2 PACKED COLUMN. Packed column air strippers are usually designed by the engineer and filled with commercially available plastic packing. The following methods are available for determining the size of a packed column air stripper: · Analytical equations: Treybal (1980), Montgomery (1985), Shulka and Hicks (1984), Ball et al. (1984). · Commercial software: "AirStrip," (Iowa State University, 1988). · Manufacturer supplied soft ware: Carbonair Environmental Systems, North East Environmental Products (see Paragraph A-5). · McCabe-Thiele graphical method: McCabe et al. (1993), Treybal (1980).

5.3.3 SIEVE TRAY. Internal components of sieve tray air strippers, such as tray dimensions, hole diameter, and weir height, are different for each manufacturer. The manufacturer or fabricator designs sieve tray strippers. Each manufacturer has software designed specifically for their units. Methods are available for determining the size of a low profile sieve tray air stripper:

5.3.4 DIFFUSED AERATION. Most "diffused aeration air strippers" are equalization basins with aeration added through diffusers to keep fine particulates in suspension. Unless the basins are extremely deep relative to their length and width, most of the

© J. Paul Guyer   2021

"stripping" is attributable to surface diffusion. This phenomenon is caused by the short contact between the air and water and the lack of turbulence compared to engineered strippers. The effectiveness of diffused air stripping has been extensively researched (Kyosai, 1991; Parker and Monteith, 1996; Sadek et al., 1996). Diffusers are designed by the diffuser manufacturer to transfer air into the water in combination with either separation of phases, as in dissolved air flotation (DAF), or mixing, as in activated sludge (AS). Water 8 (EPA, 1995) provides a model for evaluating the stripping potential of various aeration basin configurations.

5.4 TREATABILITY

5.4.1 GENERAL. Compounds with Henry's constant greater than 100 atm (moderate volatility) are generally amenable to air stripping.

5.4.2 FOULING. Retention or accumulation of solids within an air stripper is called fouling or scaling (these terms are used interchangeably). Bivalent metal ions frequently precipitate in air strippers. Influent that contains calcium (above 40 mg/L), magnesium (above 10 mg/L), iron (above 0.3 mg/L), or manganese (above 0.05 mg/L) may cause scaling (Hammer, 1975). When iron begins to precipitate, microbial growth increases the rate and amount of solids accumulation. Because the contact between air and water in any type of air stripper will result in oxidation, precautions are essential. Pre-treatment to remove interference, adding a chemical to prevent precipitation, or periodic cleaning of the air stripper to remove accumulation must be included in the design.

5.4.3 CONTAMINANT EFFECTS. Treatability studies are necessary when existing data are not adequate for predicting system performance and when interfering chemicals, such as alcohols, ketones, or surfactants, are present. A treatability study is needed in the rare instance when the concentration of the contaminant is in excess of one third of its solubility. Treatability studies are generally not required or recommended for standard air strippers operating within the range called for by the manufacturer of the trays or packing. However, if vapor pressure and solubility data for the contaminant are not available from references, such as Yaws (1994), or from the chemical manufacturer, a pilot study should be conducted.

5.4.4 LOADING RATE. A pilot study is necessary if the anticipated loading rate of either air or water is outside the range for which design information is available. Manufacturers do not recommend operation outside the loading rate limits. Low hydraulic loading may cause low mass transfer efficiency. Increased air rates over the optimum are generally a waste of energy and may decrease the rate of water flow to the point of flooding.

5.4.5 DIAMETER. Even minor variations of the stripper diameter can have a significant effect on treatability. A change in diameter results in an inverse geometric effect on the loading and distribution of both air and water phases. "Safety factors" should be completely evaluated through the calculations to assure that the over-design does not adversely affect normal operation. Large-diameter towers are not generally available for treatability studies, so geometric similarity between the pilot scale and full scale is important.

5.5 COMPARISON OF AIR STRIPPERS

5.5.1 GENERAL. The advantages and disadvantages of each type of stripper should be considered when making a selection. Either packed column or low profile air stripper will work in most situations, and are used extensively, but one may be more appropriate for the particular application. Diffused aeration strippers are less efficient for most applications, but their simplicity, ability to handle higher suspended solids, and better resistance to fouling are advantages. It may be necessary to do an economic analysis to help make the decision. Institutional factors, such as height restrictions or architectural restrictions, may require that a low profile air stripper be chosen even if a packed column stripper is more cost effective (see Table 5-1).

5.5.2 EFFICIENCY. Packed column and low profile air strippers are capable of removing more than 99% of most VOC contaminants. Increasing the depth of the packing or the number of trays will increase the stripping. Increasing the airflow through a packed column may increase the efficiency. However, increasing the airflow beyond a certain point will induce a high pressure drop and will cause flooding.

5.5.3 FOULING. Air strippers frequently become fouled by mineral deposits when calcium exceeds 40 mg/L, iron exceeds 0.3 mg/L, magnesium exceeds 10 mg/L, or manganese exceeds 0.05 mg/L, or from biological growth. Air strippers may become plugged with solids that must be removed. Packed column air strippers must either have the packing removed for cleaning or the packing must be washed with an acid solution. Both operations are time consuming and costly. Low profile air strippers are often desirable when fouling is expected. Low profile units are often fastened together, tray by tray. Small units can easily be dissembled to physically remove the biological or mineral deposits. Larger units have access ports on the side of each tray for cleaning with a high-pressure water spray. Pretreatment of the water prior to stripping is often required. Foaming control agents may be required for some liquids.

	Packed Column	Low Profile
Efficiency	Increases as packing height increases; 99%+	Increases as number of trays increases; 99%+
Cost	Lower at higher liquid flow rates	
Foam	Less foaming	
Air flow rates	Often use less air so air pollution devices, if needed, are smaller; wider range of air flow rates	
Fouling from calcium, iron, manganese suspended solids and biological growth		Easier to clean
Size	Tall	Compact; less conspicius; better appearance

Table 5-1

Comparison of Air Strippers

5.5.4 AIRFLOW RATE. The ratio of air to water flow rates is generally lower for a packed column stripper than for a low profile air stripper for the same level of VOC removal. Packed column air strippers are typically operated at 5 to 250 cfm/ft^2 (1.5 to 76 (m3/min)/m^2) of column cross-sectional area. Low profile air strippers typically operate at 30 to 60 cfm/ft^2 (9 to 18 (m^3/min)/m^2) of tray area. Thus, the tray area of a low profile air stripper will usually be much larger than the tower cross-sectional area for the same treatment conditions. Low profile units are designed to operate over a fairly narrow range of airflow rates. If the airflow rate is too high for a low profile unit, the air blowing through the trays will form a jet and disperse most of the water. This results in low removal of the VOCs. If the airflow rate is too low, the water will flow down through the holes in the sieve trays. If the water flow rate decreases to a sieve tray as the result of changed operating conditions, the airflow rate through a low profile stripper cannot be reduced correspondingly, as it will be outside the operating range specified by the manufacturer. The cost of treating the off- gas will not be decreased in proportion with the liquid loading. Packed column air strippers can operate over a wide range of airflow rates. The advantage of this is that, if the water flow rate to the column decreases, the airflow rates can also be decreased. This will reduce the cost of treating the off- gas.

5.5.5 WATER FLOW RATE. In contrast to the airflow rate, the flow rate of water through a sieve tray unit will be between 1 and 15 gpm/ft2 (0.04 to 0.6 (m3/min)/m2). Packed column strippers operate most efficiently over a narrow range of water flows, between 20 to 45 gpm/ft2 (0.8 to 1.8 (m3/min)/m2) of tower cross-sectional area (Iowa State University, 1988). The manufacturer usually designs sieve tray air strippers. Items such as the length, location, and height of the overflow weirs, weir geometry, clearance under the downcomer, fractional hold area, etc., are very important and must be designed by a manufacturer who is experienced with sieve tray columns. Additional trays can be added to many low profile air strippers if additional treatment is needed and the blower and motor are capable of handling the additional pressure drop from additional trays. Combining the airflow rate and the water flow rate results in an air-to-water ratio as low as 30 to as high as several hundred (volume to volume) for sieve tray units.

5.5.6 PRESSURE DROP AND POWER CONSUMPTION. The pressure drop through the packing of a packed tower air stripper is often lower than the pressure drop through a comparable low profile unit. This allows a smaller blower and motor, with reduced electrical operating costs.

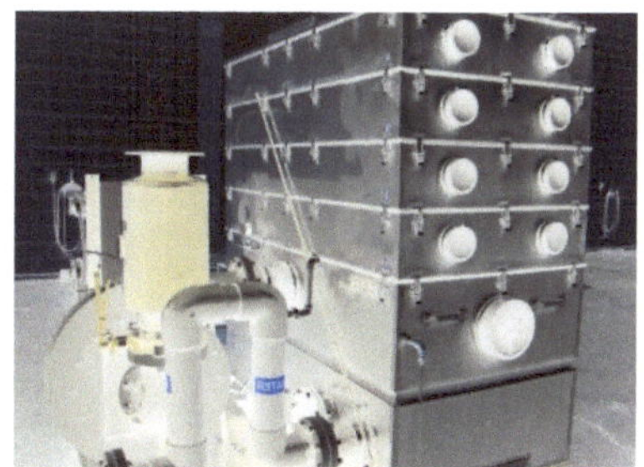

5.6 AIR POLLUTION CONTROL REQUIREMENTS

5.6.1 GENERAL. Off-gas from an air stripper may or may not need to be controlled, depending on the level of contaminants and on state and local regulations. If the loading of a contaminant or contaminants exceeds the amount allowed by regulation to be discharged to the air, treatment of the stripper off- gas will be required. This must be thoroughly evaluated, as air pollution controls will add major capital and O&M costs to the system. Packed columns use less air for a given loading of water and contaminant than do sieve tray air strippers. This is important when air pollution regulations require that the air leaving the unit be treated to remove the volatile organic chemicals before it is discharged to the atmosphere. In these cases, achieving the lowest airflow rate, and in turn the lowest air pollution control costs, may be the driving force in determining which type of air stripper to use.

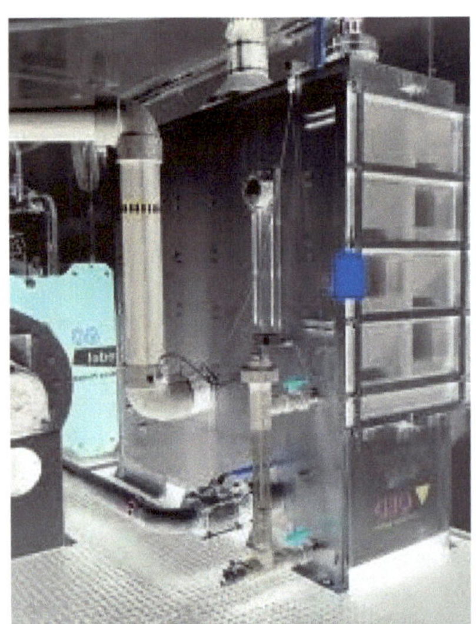

$$x_{ai} L + y_{ai} G = x_{ae} L + y_{ae} G$$

which is the flow rate of contaminant a, where

$$L \quad = \text{ molar flow of liquid (water)}$$
$$G \quad = \text{ molar flow of gas (air)}$$
$$x_{ai} = \text{ mole fraction of contaminant } a \text{ in influent liquid (water)}$$
$$x_{ae} = \text{ mole fraction of contaminant } a \text{ in effluent liquid (water)}$$
$$y_{ai} = \text{ mole fraction of contaminant } a \text{ in influent gas (air)}$$
$$y_{ae} = \text{ mole fraction of contaminant } a \text{ in effluent gas (air)}.$$

Assuming an uncontaminated air supply, we find:

$$y_{ai} = 0$$
$$x_{ai} L = x_{ae} L + y_{ae} G$$

Rearranging terms yields:

$$y_{ae} G = \left(x_{ai} - x_{ae} \right) L$$

$$\frac{G}{L} = \left(\frac{x_{ai} - x_{ae}}{y_{ae}} \right)$$

$$\% \text{ Removal} = \left(\frac{x_{ai} - x_{ae}}{x_{ai}} \right) 100\%$$

where

$$p_{Te} = \text{ total pressure of gas(air) effluent}$$
$$p_{ae} = \text{ partial pressure of contaminant in a gas (air) effluent}$$

From Dalton's Law of partial pressures,

$$y_{ae} = \frac{p_{ae}}{p_{Te}} \frac{mole}{mole}$$

At equilibrium, from Henry's law:

$$p_{ae} = H_a\, x_{ai}\ \text{atm}$$

Substituting for p_{ae} yields:

$$y_{ae}\, p_{Te} = H_a\, x_{ai}\ \text{atm}$$

$$y_{ae} = \frac{H_a\, x_{ai}}{p_{Te}}\ \text{atm}$$

For the air pollution worst case, it must be assumed that all volatile contaminants introduced to the stripper are transferred to the air. The only positive control of pollutant transfer to the air is the rate of contaminated water pumped to the stripper.

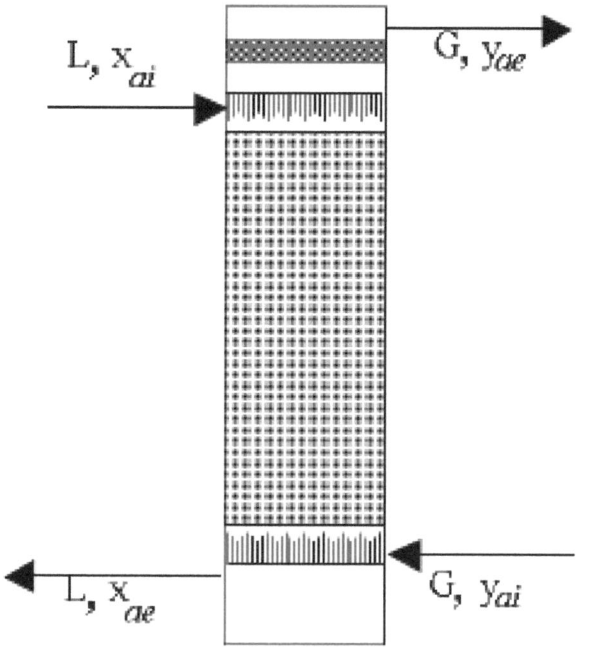

Figure 6-1

Material Balance.

5.6.2 OFF-GAS TREATMENT. Activated carbon and thermal oxidation are commonly used to treat the off- gas. The unit efficiency of either method is directly proportional to the concentration of the contaminant in the off- gas. Activated carbon is simple but does not destroy the contaminant and may result in potential disposal costs. Thermal oxidation destroys the contaminant but is more complex. Air pollution control devices should be evaluated before determining which device to use. "As can be expected, the lower the concentration [of VOCs] in the gas stream the higher the control cost".

5.6.3 INNOVATIVE AIR POLLUTION CONTROL DEVICES. Information on innovative air pollution control devices can be found in the Remediation Technologies Screening Matrix and Reference Guide.

5.7 FLOODING

5.7.1 GENERAL. Air strippers depend on a balance between air and water to provide the necessary intimate contact within the stripper. The induced turbulence within the system provides the energy required for separation of the volatile organics from the water. The air carries the volatiles away from the water.

5.

7.2 OCCURENCE OF FLOODING. Excess airflow will cause flooding in air strippers, regardless of type. Extra blower head requirements (i.e., for "future" additional trays) should not be "thrown in" as a safety factor without providing an orifice plate to "burn up" the excess head. Obviously, if this is done, energy is wasted in oversizing.

5.7.3 BLOWER. A centrifugal blower can be modified to reduce the air output by changing or trimming the impellers. Impeller must be trimmed properly to maintain blower balance. Changing the blower sheaves and belts can change the speed and corresponding output of belt driven blowers.

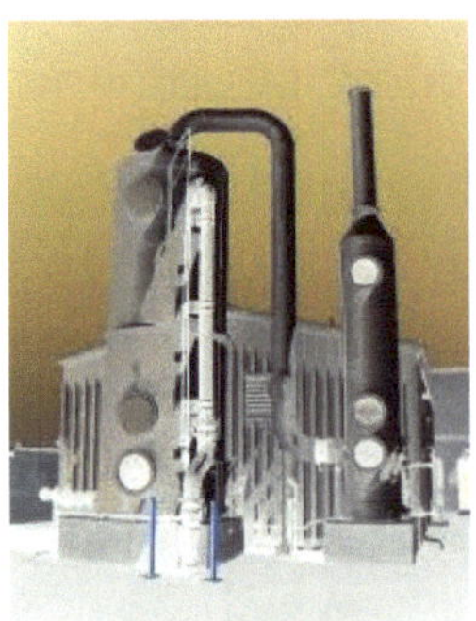

5.8 PROCESS CONTROL

5.8.1 GENERAL. The pumps delivering the water and the blowers delivering the air are packaged with contacts, controllers, and appropriate alarms.

5.8.2 LEVEL CONTROLS. If the plant hydraulics or sampling requirements mandate that the effluent sump cannot overflow, it must be equipped with level control and level alarms to prevent this. Feedback from the effluent sump level should turn the well or influent pumps down or off and activate an alarm when the sump level setting is exceeded. The controller for effluent pumps should allow alternating operation of lead, lag, and stand-by pumps, with low and high level alarms and pump control over-ride functions.

5.8.3 PRESSURE CONTROLS. Feedback from the air stripper pressure sensors mounted in influent and effluent piping should turn down or shut off the blower and activate the alarm when the differential pressure across the stripper begins to rise. For energy economy, the blower control should also be interlocked with water flow to the stripper. The pressure differential disappears if the blower fails.

5.9 ECONOMIC EVALUATION. Consider the advantages and disadvantages of each type of air stripper, along with capital costs, installation costs, and long term O&M, when the stripping system is designed. Remedial Action Cost Engineering and Requirements System (RACER, Paragraph A-5) software can be used to compare costs. Costs for treating off- gas and fouling are major factors in the economic evaluation. Architectural restrictions may require a low profile air stripper or multiple strippers in series, even if a single packed tower stripper would, otherwise, be more cost effective. See Figure 9-1 for the cost model.

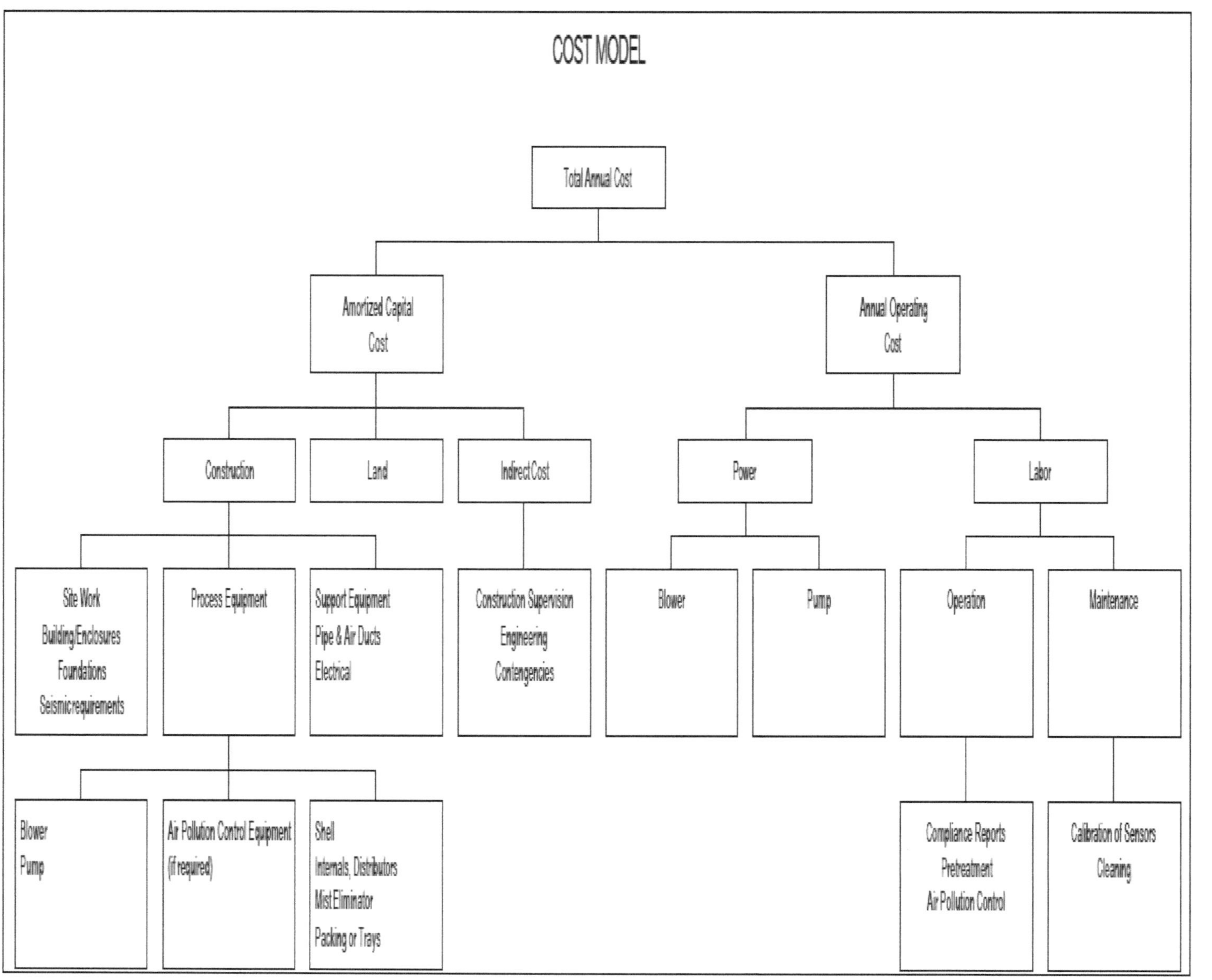

Figure 9-1

Cost model.

5.10 EXAMPLES OF AIR STRIPPING BY LOW PROFILESIEVE TRAY DEVICE

C-1. Example in SI Units. This example will illustrate a method of making preliminary design calculations to size a low profile sieve tray air stripper. Final designs depend heavily on the design of the trays. Unfortunately, this information is often not available to the designer. As a result, the final design and size of the unit must be determined from information supplied by the manufacturer. Low profile sieve tray air strippers are usually secured as complete units assembled on skids at the factory and shipped as a unit rather than being designed and constructed from job drawings and specifications. The steps in the preliminary design calculations follow (refer to Figure C-1).

- Determine the minimum and maximum volume of water to be air stripped, the minimum temperature of the water, and the maximum concentration of volatile organic chemicals (VOC) in the untreated water to be air stripped.

- Determine the desired concentration (percent removed) of the VOC in the treated water.

- Calculate the theoretical number of sieve trays needed to remove the VOC to the desired concentration.

- Estimate the tray efficiency and the number of actual trays needed.

- Estimate the size (cross-sectional area) of the perforated plate section of each tray.

- Estimate the pressure drop through the air stripper.

- Estimate the size of the air blower motor (kW).

a. Determine the volume of water to be air stripped, the minimum temperature of the water, and concentration of all the volatile organic chemicals (VOC) in the untreated water. The inlet water contains 10 mg/L of the volatile organic chemical (VOC) trichloroethylene (TCE). (Note: If the inlet water contains more than one VOC, repeat the process for each to estimate the number of trays needed for each VOC. Use the largest number of trays for the estimated design.) The flow rate of water to be treated is 0.2 m^3 per minute. The minimum temperature of the water is 20°C.

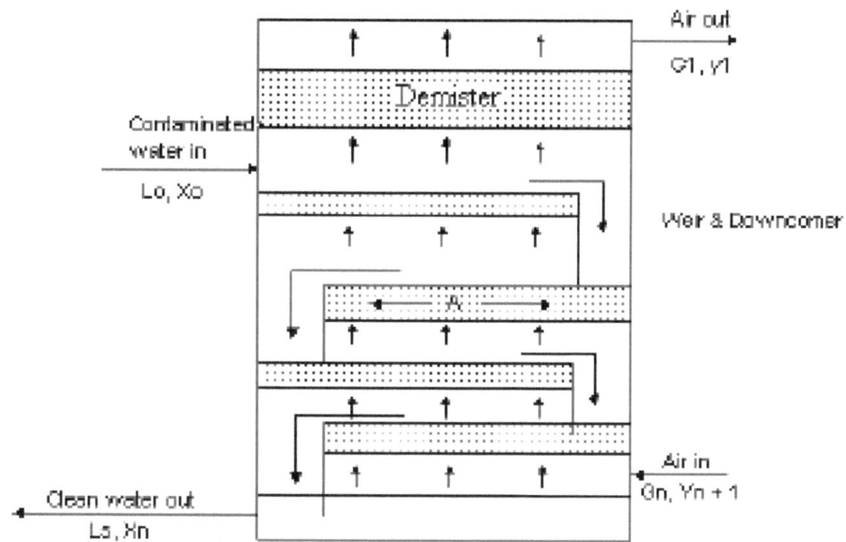

Figure C-1. Cross-sectional area of perforated plate section.

b. Determine the desired concentration of the TCE in the treated water. The desired concentration of TCE in the discharge water is 0.1 mg/L (99% removal).

c. Calculate the theoretical number of sieve trays needed to remove the VOC to the desired concentration. The theoretical number of trays required is estimated by using the following relationship (Treybal, 1980):

$$N_{theoretical} = \frac{\log\left[\dfrac{X_0 - \dfrac{Y_{n+1}}{m}}{X_n - \dfrac{Y_{n+1}}{m}}\left(1 - \dfrac{1}{S}\right) + \dfrac{1}{S}\right]}{\log S}$$

where

X_0 = concentration of contaminant (TCE) in the inlet water phase: 10 mg/L
X_n = concentration of contaminant (TCE) in the treated water phase: 0.1 mg/L

N = number of theoretical plates. Assumes that the liquid on each plate is completely mixed and that the vapor leaving the plates is in equilibrium with the liquid.

H = Henry's constant (kPa)

m = slope of equilibrium curve (H/Pt)

G = kg-moles air/min

L = kg-moles of water/min

Pt = ambient pressure (kPa)

S = stripping factor (mG/L)

Y_{n+1} = concentration of volatiles in the air entering the air stripper.

(1) For air stripping $Y_{n+1} = 0$ (the concentration of TCE in the air entering the air stripper is zero) and the equation becomes:

$$N_{theoretical} = \frac{\log\left[\frac{(X_0)}{(X_n)}\left(1-\frac{1}{S}\right)+\frac{1}{S}\right]}{\log\ S}$$

(2) In this example, the inlet concentration of TCE (X_0), the desired outlet concentration of TCE (Xn), the liquid temperature, and flow rate are known. The airflow rate (G) must be determined and is related to the perforated plate area of each tray. Several combinations of airflow rates and number of trays should be calculated to determine the best economic balance between having more trays and a lower airflow rate (higher capital costs vs. lower operating costs) and fewer trays and a higher air flow rate (lower capital costs vs. higher operating costs). An economic comparison is beyond the scope of this example. For this example use an air-to-water ratio of 37 m³ of air to 1 m³ of water (see paragraph 5-5).

(3) Substituting into the above equation yields:

$H = 5.57 \times 10^4$ kPa (for TCE at 20°C)

$Pt = 101$ kPa (101 kPa at sea level, 86 kPa at 1500 m elevation)

$$m = \frac{H}{Pt} = \left(\frac{5.57 \times 10^4\ kPa}{101 kPa}\right) = 551\ \frac{mole\ H_2O}{mole} = \frac{551 kgmole H_2O}{kgmole\ air}$$

$$G = \frac{37 m^3\ air}{min} \times \frac{kg-mole\ air}{24.0 m^3\ air\ @\ 20°C} = 1.54\ \frac{kg-mole\ air}{min}$$

© J. Paul Guyer   2021   150

$$L = \frac{1m^3\ H_2O}{min} \times \frac{1000kg\,H_2O}{m^3\ H_2O} \times \frac{kgmole\,H_2O}{18kg\ H_2O} = \frac{55.6\ kg-moleH_2O}{min}$$

$$S = \left(\frac{551kg\ mole\ H_2O}{kgmoleair}\right)\left(\frac{1.54kgmole\ \dfrac{air}{min}}{55.6kgmole\ \dfrac{water}{min}}\right)$$

$$= 15.3$$

$$N_{theor} = \frac{\log\left[\dfrac{10\dfrac{mg}{L}}{0.1\dfrac{mg}{L}} \times \left(1-\dfrac{1}{15.3}\right) + \dfrac{1}{15.3}\right]}{\log 15.3} = 1.66$$

d. Estimate the tray efficiency and the number of actual trays needed. In actual practice a condition of complete equilibrium does not exist. The overall plate efficiency is:

$$E = \frac{N_{theoretical}}{N_{actual}}$$

Rearranging gives the number of actual trays as:

$$N_{actual} = \frac{N_{theoretical}}{E}$$

The efficiency is highly dependent on the design of the trays and the vapor flow rate. From manufacturer's data, the appropriate range appears to be $E = 0.4$ to 0.6 (i.e., 40 to 60% efficient). Using the above relationship and assuming 50% tray efficiency, and substituting into the above equation, gives the number of actual trays needed as:

$$N_{actual} = \frac{1.66}{0.50} = 3.32 = 4$$

e. Estimate the size (cross-sectional area) of the perforated plate section of each tray. The cross-sectional area of the perforated plate section of each tray is related to the airflow rate; 9 to 18 m^3 per minute per m^2 of tray area is common (see Paragraph 5-4). For this example, use 18 m^3 per minute per m^3 of tray area. The area is:

$$\frac{0.2\,m^3\,H_2O}{min} \times \frac{37\,m^3\,air}{m^3\,H_2O} \times \frac{m^2\,plate\,area}{18\,m^3\,\frac{air}{min}} = 0.41m^2$$

Estimate that the downcomer and weir area is 20% of each plate. The total cross-sectional area of each plate is:

$$0.41m^2 + 0.41 \times 0.2m^2 = 0.49\ m^2$$

f. Estimate the pressure drop through the air stripper. Most of the pressure drop through the air stripper is from the head of liquid on each tray times the number of trays. The depth of liquid on the trays typically varies from 8 to 12 cm of water. Assume 10 cm water for this example. The other pressure drop is from the piping from the blower to the air stripper and inlet and exit losses in the column. This will vary from system to system. An estimate for these losses is 25 cm water. From this information, the total pressure drop through the system is as follows:

$$N\ trays = 3.32;\ round\ up\ to\ 4$$

$$4\ trays \times 10cm\ \frac{H_2O}{tray} + 25cm\ H_2O = 65cm\ H_2O$$

g. Estimate the size of the blower motor (kW). The size of the blower motor is a function of the flow rate of air and the pressure drop. Methods of estimating the size of the blower motor can be found in reference books (McCabe et al., 1993; Avallone and Baumeister, 1987; Perry, 1984) and will not be calculated in this example.

C-2. Example in English Units. This example will illustrate a method of making preliminary design calculations to size a low profile sieve tray air stripper. Final designs depend heavily on the design of the trays. Unfortunately, this information is often not available to the designer. As a result, the final design and size of the unit must be determined from information supplied by the manufacturer. Low profile sieve tray air strippers are usually secured as complete units assembled on skids at the factory and shipped as a unit rather than being designed and constructed from job drawings and specifications. The steps in the preliminary design calculations follow (refer to Figure C-1).

- Determine the minimum and maximum volume of water to be air stripped, the minimum temperature of the water, and the maximum concentration of volatile organic chemicals (VOC) in the untreated water to be air stripped.

- Determine the desired concentration (percent removed) of the VOC in the treated water.

- Calculate the theoretical number of sieve trays needed to remove the VOC to the desired concentration.

- Estimate the tray efficiency and the number of actual trays needed.

- Estimate the size (cross-sectional area) of the perforated plate section of each tray.

- Estimate the pressure drop through the air stripper.

- Estimate the size of the blower motor (hp).

a. Determine the volume of water to be air stripped, the minimum temperature of the water and concentration of all the volatile organic chemicals (VOC) in the untreated water. The inlet water contains 10 mg/L of the volatile organic chemical (VOC) trichloroethylene (TCE). (Note: If the inlet water contains more than one VOC, repeat the process for each to estimate the number of trays needed for each VOC. Use the largest number of trays for the estimated design.) The flow rate of water to be treated is 50 gpm. The minimum temperature of the water is 60°F.

b. Determine the desired concentration of the TCE in the treated water. The desired concentration of TCE in the discharge water is 0.1 mg/L (99% removal).

c. Calculate the theoretical number of sieve trays needed to remove the VOC to the desired concentration. The theoretical number of trays required is estimated by using the following relationship (Treybal, 1980):

$$N_{theoretical} = \frac{\log\left[\left(\dfrac{X_0 - \dfrac{Y_{n+1}}{m}}{X_n - \dfrac{Y_{n+1}}{m}}\right)\left(1 - \dfrac{1}{S}\right) + \dfrac{1}{S}\right]}{\log S}$$

where

X_0	=	concentration of contaminant (TCE) in the inlet water phase: 10 mg/L
X_n	=	concentration of contaminant (TCE) in the treated water phase: 0.1 mg/L
N	=	number of theoretical plates. Assumes that the liquid on each plate is completely mixed and that the vapor leaving the plates is in equilibrium with the liquid.
H	=	Henry's Constant (atm)
m	=	slope of equilibrium curve (H/Pt)

$$G = \text{lb-moles air/min}$$
$$L = \text{lb-moles of water/min}$$
$$S = \text{stripping factor } (mG/L)$$
$$Pt = \text{ambient pressure (atm)}$$
$$Y_{n+1} = \text{concentration of volatiles in the air entering the air stripper.}$$

(1) For air stripping $Y_{n+1} = 0$ (the concentration of TCE in the air entering the air stripper is zero) and the equation becomes:

$$N_{theoretical} = \frac{\log\left[\left(\frac{X_0}{X_n}\right)\left(1 - \frac{1}{S}\right) + \frac{1}{S}\right]}{\log S}$$

(2) In this example, the inlet concentration of TCE (X_0), the desired outlet concentration of TCE (Xn), and the liquid temperature and flow rate are known. The airflow rate (G) must be determined and is related to the perforated plate area of each tray. Several combinations of air flow rates and number of trays should be calculated to determine the best economic balance between having more trays and a lower air flow rate (higher capital costs vs. lower operating costs) and fewer trays and a higher air flow rate (lower capital costs vs. higher operating costs). An economic comparison is beyond the scope of this example. For this example use an air-to-water ratio of 5 cfm of air to 1 gpm of water.

(3) Substituting into the above equation yields (for TCE at 20°C):

$$H = 550\,\text{atm}$$

$$Pt = 1.0\,\text{atm}\,(\text{Note}:1.0\,\text{atm at sea level}, 0.86\,\text{atm at }5000\,\text{ft})$$

$$m = \frac{H}{Pt} = \frac{550\,\text{atm}}{1\,\text{atm}} = \frac{550\ \text{lb-mole}\,H_2O}{\text{lb-mole air}}$$

$$G = \frac{5\,\text{ft}^3}{\text{min}} \times \frac{\text{lb}-\text{mole air}}{380\,\text{ft air}\left(@60°F\right)} = 0.0132\ \frac{\text{lb}-\text{mole air}}{\text{min}}$$

$$L = \frac{1\,\text{gal}}{\text{min}} \times \frac{8.34\,\text{lb}\,H_2O}{\text{gal}\,H_2O} \times \frac{\text{lb}-\text{mole}\,H_2O}{18\,\text{lb}} = \frac{0.463\,\text{lb}-\text{mole}\,H_2O}{\text{min}}$$

$$S = \frac{550 \ \text{lb-mole } H_2O}{1 \text{lb-mole air}} \times \frac{0.0132 \text{lb} - \text{mole/min}}{0.463 \text{lb} - \text{mole/min}} = 15.7$$

$$N_{theor} = \frac{\log\left[\dfrac{10\dfrac{\text{mg}}{\text{L}}}{0.1\dfrac{\text{mg}}{\text{L}}} \times \left(1 - \dfrac{1}{15.7}\right) + \dfrac{1}{15.7}\right]}{\log 15.7} = 1.65$$

d. Estimate the tray efficiency and the number of actual trays needed. In actual practice, a condition of complete equilibrium does not exist. The overall plate efficiency is:

$$E = \frac{N_{theoretical}}{N_{actual}}$$

Rearranging gives the number of actual trays as:

$$N_{actual} = \frac{N_{theoretical}}{E}$$

The efficiency is highly dependent on the design of the trays and the vapor flow rate. From manufacturer's data, the appropriate range appears to be $E = 0.4$ to 0.6 (i.e. 40 to 60% efficient). Using the above relationship and assuming 50% tray efficiency, and substituting into the above equation, gives the number of actual trays needed as:

$$N_{actual} = \frac{1.65}{0.50} = 3.3 = 4$$

e. Estimate the size (cross-sectional area) of the perforated plate section of each tray. The cross-sectional area of the perforated plate section of each tray is related to the airflow rate; 30 to 60 cfm/ft^2 is common (see Paragraph 5-4) For this example, use 60 cfm/ft^2. Using this and the air-to-water ratio of 5 cfm of air to 1 gpm of water and the water flowrate of 50 gpm gives the cross-sectional area as

$$50\,\text{gpm} \times \frac{5\,\text{cfm}}{1\,\text{gpm}} = 250\ \text{cfm}$$

$$250\,\text{cfm} \times \frac{1\text{ft}^2}{60\,\text{cfm}} = 4.17\text{ft}^2$$

Estimate that the downcomer and weir area is 20% of each plate. The total cross-sectional area of each plate is:

$$4.17 + 4.17 \times 0.2 = 5.0\text{ft}^2$$

f. Estimate the pressure drop through the air stripper. Most of the pressure drop through the air stripper is from the head of liquid on each tray times the number of trays. The depth of liquid on the trays typically varies from 3 to 5 in. of water. Assume 4 in. for this example. The pressure drop from the air flowing through the holes in the sieve tray is usually insignificant to the other pressure drops in the system and will be ignored for this example. The other pressure drop is from the piping from the blower to the air stripper and inlet and exit losses in the column. This will vary from system to system. An estimate for these losses is 10 in. From this information the total pressure drop through the system is as follows:

$$N\ \text{trays} = 3.3;\ \text{round up to 4}$$

$$4\ \text{trays} \times 4\text{in.wg} \, \frac{H_2O}{\text{tray}} + 10\text{in.wgH}_2O = 26\text{in.wgH}O$$

g. Estimate the size of the blower motor (hp). The size of the blower motor is a function of the flow rate of air and the pressure drop. Methods of estimating the size of the blower motor (hp) can be found in reference books (McCabe et al., 1993; Avallone and Baumeister, 1987; Perry, 1984) and will not be calculated in this example.

5.11 EXAMPLE OF AIR STRIPPING BY PACKED COLUMN

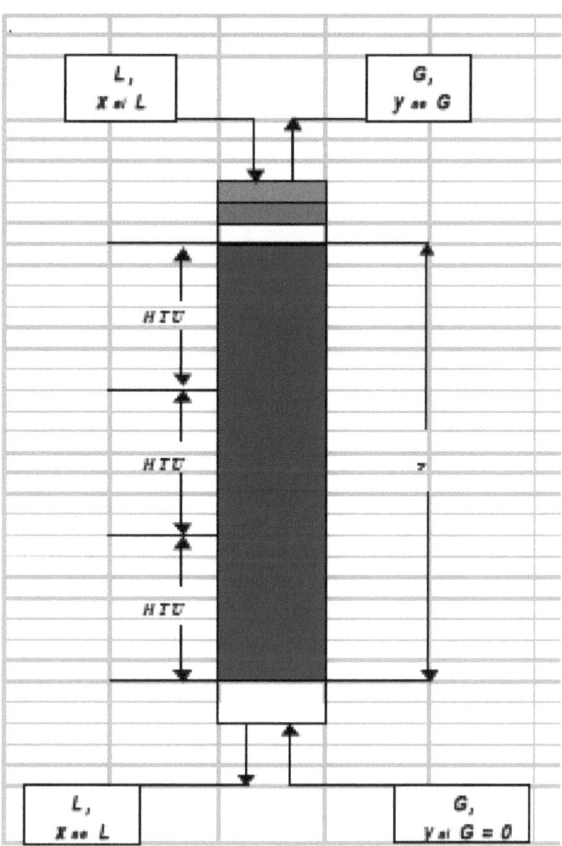

Figure D-1. Random "dumped" packed tower.

D-1. Parameters.

NTU = number of transfer units

HTU = height of transfer unit [m]

Z = $NTU \times HTU$ packing depth [m]

L = molar liquid (water) flow per unit of stripper cross-sectional area $\left[\dfrac{\text{kg-mole}}{\text{m}^2 \text{ sec}} \right]$

G = molar gas (air) flow per unit of stripper cross-sectional area $\left[\dfrac{\text{kg-mole}}{\text{m}^2 \text{ sec}} \right]$

x_{ai} = mole fraction of contaminant a in liquid (water) influent $\left[\dfrac{\text{kg-mole}}{\text{kg-mole water}} \right]$

$$x_{ae} = \text{molefraction of contaminant } a \text{ in liquid(water) effluent} \left[\frac{\text{kg-mole}}{\text{kg-mole water}} \right]$$

$$y_{ai} = \text{molefraction of contaminant } a \text{ in gas(air) influent} \left[\frac{\text{kg-mole}}{\text{kg-mole air}} \right]$$

$$y_{ae} = \text{molefraction of contaminant } a \text{ in gas(air) effluent} \left[\frac{\text{kg-mole}}{\text{kg-mole air}} \right]$$

$$(x_{ai} - x_{ae}) L = (y_{ae} - y_{ai}) G$$

which is moles of contaminant a transferred from liquid to gas per unit of stripper cross-sectional area per unit time (kg-mole/s)

$$\left(\frac{x_{ai} - x_{ae}}{y_{ae} - y_{ai}} \right) = \frac{G}{L} \left[\frac{\text{kg-mole air}}{\text{kg-mole water}} \right]$$

which is the molar ration of gas (air) to liquid (water), and assumining uncontaminated influent air:

$$y_{ai} = 0$$

$$\frac{G}{L} = \left(\frac{x_{ai} - x_{ae}}{y_{ae}} \right)$$

where x_{ai} and L are field measurements and x_{ae} is imposed by ARAR, and

$$p_{Te} = \text{total pressure of gas (air) effluent (atm)}$$
$$P_{ae} = \text{partial pressure of contaminant } a \text{ in gas (air) effluent (atm).}$$

From Dalton's Law of partial pressures:

$$y_{ae} = \frac{P_{ae}}{p_{Te}} \left[\frac{\text{mole}}{\text{mole}} \right] \text{or} \left[\frac{\text{atm}}{\text{atm}} \right]$$

$$P_{ae} = y_{ae}\, p_{Te} \left[\text{atm} \right]$$

at equilibrium from Henry's Law:

$$P_{ae} = H_a\, x_{ai} \left[\text{atm} \right]$$

substituting yields:

$$y_{ao} \, p_{To} = H_a \, x_{ai} \; [atm]$$

$$y_{ao} = \frac{H_a \, x_{ai}}{p_{To}} \; [atm]$$

and from the material balance:

$$\left(x_{ai} - x_{ao} \right) \left(\frac{L}{G} \right) = y_{ao} \left[\frac{mole}{mole} \right]$$

Again substituting gives

$$\left(x_{ai} - x_{ao} \right) \left(\frac{L}{G} \right) = \frac{H_a \, x_{ai}}{p_{To}}$$

$$\frac{x_{ai} - x_{ao}}{x_{ai}} = \frac{H_a \left(\dfrac{G}{L} \right)}{p_{To}}$$

The fraction of contaminant transferred from liquid (water) to gas (air) phase is:

$$\frac{C_{ai} - C_{ao}}{C_{ai}} = \frac{x_{ai} - x_{ao}}{x_{ai}}$$

where

C_{ai} = concentration of contaminant a in liquid (water) influent [μg/L]

C_{ae} = concentration of contaminant a in liquid (water) effluent [μg/L].

For convenience, the flows of water and air are measured volumetrically

$$C_{ai} (L) \, Q_L = C_{ao} (L) \, Q_L + C_{ao} (G) \, Q_G$$

and

$$\frac{C_{ai} - C_{ao}}{C_{ai}} = \left(\frac{H_a}{p_{To}} \right) \left(\frac{Q_G}{Q_L} \right)$$

where p_{Te} is measured as a fraction of the standard atmosphere (atm), H'_a is the dimensionless Henry's constant H_a/C_0RT, actually (volume/volume), Q_g/Q_L is reduced to common flow units $[m^3/m^3]$, and C_0 is the molar density of water at 20°C, 55.41 kg mole/m^3. The theoretical minimum, equilibrium, moles of gas required G_{min}/L is calculated from the influent and effluent concentrations and the "dimensionless" Henry's constant (H_a).

$$R_u \quad = \quad 0.08205746 \left[\frac{m^3\ atm}{kg-mole\ K} \right] \quad \text{the universal gas constant}$$

At 1 atm and 20°C the molar density of water is C_0, 55.41 kg-mole/m^3. Q_G/Q_L $[m^3/m^3]$ is the air-to-water ratio, ATW.

$$y_{ae} = H_a\, x_{ai}/p_{Te}\ \text{(mole/mole)}$$

Substituting gives

$$\frac{L\left(x_{ai} - x_{ae}\right)}{G} = \frac{H_a\, x_{ai}}{p_{Te}}$$

and rearranging yields

$$\frac{G_{min}}{L} = \frac{\left(x_{ai} - x_{ae}\right) p_{Te}}{H_a\, x_{ai}}$$

which is the equilibrium molar ratio of gas (air) to liquid (water).

D-2. Develop the Design Basis.

a. Characterize the influent conditions and effluent requirements, including RI/FS data + total organics + background inorganics and minimum water temperature.

© J. Paul Guyer   2021   161

Table D-1
Contaminants

Contaminant	Formula	GMW* [g/g-mole]	CAS Number	H_a** [atm/mole/mole]
Benzene	C_6H_6	78.11	71-43-2	309.2
Toluene	$C_6H_5CH_3$	92.14	108-88-3	353.1
Trichloroethylene (TCE)	C_2HCl_3	131.50	79-01-6	506.1

*The [gram] molecular weight of the contaminant.
** H_a at 20°C (296.13 K).

   b. Design the pumping system to maintain the flow. Use the real flow rate, not rounding up. Discharge head adjustments for the stripper are added to the TDH. The aggregate flow from the hydraulic barrier is 440 gpm (0.0278 m^3/s) in this example.

   c. Design the pre-treatment system to prevent scale/slime from clogging the stripper (if water is high in hardness, iron or manganese).

Table D-2
Background Inorganic Concentrations

Ion	mgL	GMW	Valence	GEqW*	meq/L	mg/L as$CaCO_3$
CO_2	0	44	−2	22	0.00	0.00
				Anions		
SO_4	60	96	−2	48	1.25	62.46
Cl	54	35	−1	35	1.52	76.15
HCO_3	30	61	−1	61	0.49	24.58
					TOTAL	163.19
$CaCO_3$		100	0	50	0.00	0.00
				Cations		
Na	10	23	1	23	0.43	21.75
Ca	40	40	2	20	2.00	99.80
Fe	0.3	56	2	28	0.01	0.54
Mg	10	24	2	12	0.82	41.12
Mn	0.05	55	2	27	0.00	0.09
					TOTAL	163.29

* GEqW is the [gram] equivalent weight of the inorganic ion.

d. Construct a contaminant material balance for the stripping system.

Table D-3
Removal Requirements

| Contaminant | Concentration [μg/L] | | | Mole Fraction [mole/mole] | |
	Influent, C_{ai}	Effluent Standard, C_{ae}	Removal Requirement	x_{ai}	x_{ae}
Total VOCs	2500	NA	NA	NA	NA
Benzene	750	10	98.7%	0.17330	0.00231
Toluene	1000	100	90.0%	0.19588	0.01959
Trichloroethylene (TCE)	750	100	86.7%	0.10294	0.01373

e. Assess the air pollution control requirements from the material balance and the regulations.

D-3. Determine the Column Diameter.

a. Determine a preliminary stripper cross-sectional area for the sustained pumping rate, 440 gpm (0.02776 m³/s) using 45 gpm/ft² (0.03056 m/s) for the stripper surface loading.

$$A = \frac{Q\ \text{ft}^2}{45\,\text{gpm}} \left(\frac{Q\ \text{m}^2\,\text{s}}{0.03056\ \text{m}^3} \right)$$

$$= 0.0222 \frac{Q\,\text{ft}^2}{\text{gpm}} \left(32.72\,Q\,\frac{\text{s}}{\text{m}} \right)$$

$$= 0.0222(440\,\text{gpm})\ \frac{\text{ft}^2}{\text{gpm}} \left(32.72 \left(0.02776\ \frac{\text{m}^3}{\text{s}} \right) \frac{\text{s}}{\text{m}} \right)$$

$$= 9.7778\,\text{ft}^2 \left(0.9084\ \text{m}^2 \right)$$

b. Divide the are by the number of strippers.

$$a = \frac{A}{\#}$$

$$= \frac{9.7778}{2}\,ft^2 \left(\frac{0.9084m^2}{2}\right)$$

$$= 4.889\,\frac{ft^2}{stripper} \left(\frac{0.4542m^2}{stripper}\right)$$

c. Divide $a = \pi\,(d^2/4)$ the unit area by π, multiply by 4 and take the square root.

$$d \approx \sqrt{\frac{4a}{\delta}}$$

$$d \approx \sqrt{\frac{4\,(4.889)}{\delta}} \left(\sqrt{\frac{4\,(0.4542)}{\delta}}\right)$$

$$d \approx \sqrt{6.22473}\;(\sqrt{0.5783})$$

$$d \approx 2.5ft\;(0.762m)$$

d. Bracket the calculated diameter with the nearest standard diameters. In this example, a 2.5-ft (0.762-m) diameter column is standard for most manufacturers. The availability of standard metric sizes should be verified.

D-4. Find a Suitable Packing.

a. Find packings in the diameter range of roughly 5 to 10% of the stripper diameter. The rule of thumb is 1 in. of packing diameter per 1 ft of tower diameter; 2.5 in. (0.0635 m) packing is not standard for most manufacturers.

b. Reconsider the number of strippers if the packings and diameters don't correspond. Three 2-ft diameter strippers with 2-in. packing could be used in lieu of two 2.5-ft-diameter strippers.

b. Find the area of the standard diameter strippers.

$$a = \delta\,\frac{d^2}{4}$$

$$= \frac{(2.5ft\delta)^2}{4} \left(\frac{(0.762\ m)\delta^2}{4}\right)$$

$$= 4.908ft^2\,(0.456m^2)$$

d. Calculate the surface hydraulic loading Q/A and compare the loading with various packing manufacturers' recommendations.

$$\frac{Q_L}{A} = \frac{220\text{gpm}}{4.908\text{ft}^2}\left(\frac{0.01388\frac{m^3}{s}}{0.456\,m^2}\right) \text{ per stripper}$$

$$V_L = 44.82\,\frac{\text{gpm}}{\text{ft}^2}\left(0.03044\,\frac{m}{s}\right)$$

e. Adjust the system configuration to get the hydraulics within the recommended range.

D-5. Calculate the Minimum Gas Flow. Determine G_{min} and the critical contaminant from the following relationship:

$$\frac{Q_{G\,min}}{Q_L} = \frac{(C_{ai} - C_{ae})}{H_a C_{ai}}$$

Table D-4
Critical Contaminant

For P_{te} = 1 atm and 20°C (296.13 K) $H_a = H_a/C_o\,R\,T$		
Contaminant	$\dfrac{(C_{ai} - C_{ae})}{C_{ai}}$	H_a	$\dfrac{Q_{G\,min}}{Q_L}$
Benzene	0.9867	0.2320	$4.253\,\dfrac{m^3}{m^3}$
Toluene	0.9000	0.2649	$3.397\,\dfrac{m^3}{m^3}$
Trichloroethylene (TCE)	0.8667	0.3797	$2.283\,\dfrac{m^3}{m^3}$
Critical Contaminant (Benzene)		
$\dfrac{Q_{G\,min}}{Q_L}$	=	$4.253\,\dfrac{m^3}{m^3}$ (maximum)

D-6. Calculate the Mass Transfer Rate. Use a model, if available, to confirm the results.

$$\frac{a_w}{a_t} = 1 - e^{-1.45\left(\frac{s_c}{s}\right)^{0.75} N_{Re}^{\,0.01} N_{Fr}^{\,-0.05} N_{we}^{\,0.2}}$$

$$\frac{1}{K_{LA}} = \frac{1}{H_a K_G a_w} + \frac{1}{K_L a_w}$$

where

$$
\begin{array}{lll}
a_w & = & \text{wetted surface area of the packing}(m^2/m^3) \\
a_t & = & \text{total surface area of the packing } (m^2/m^2) \\
K_{LA} & = & \text{overall mass transfer rate (m/s)} \\
K_L & = & \text{liquid phase mass transfer rate (m/s)} \\
K_G & = & \text{gas phase mass transfer rate (m/s).}
\end{array}
$$

a. Calculate the dimensionless numbers (http://www.processassociates.com/process/dimen gives a comprehensive listing and definitions of dimensionless numbers).

$$N_{Re} = \left(\frac{1}{a_t}\right)\frac{V_L \rho_L}{\mu_L} \qquad \text{Reynolds Number}$$

$$N_{Fr} = a_t \frac{V_L^2}{g_c} \qquad \text{Froude Number}$$

$$N_{We} = \left(\frac{1}{a_t}\right)\frac{V_L^2 \rho_L}{g_c S} \qquad \text{Weber Number}$$

$$N_{Sc} = \frac{\mu_L}{\rho_L D_L} \qquad \text{Schmidt Number}$$

$$g_c = 9.807 \frac{m}{s^2} \qquad \text{gravitation constant}$$

b. Look up the properties of the liquid (water) at the minimum water temperature, T (Table D-5)

$$s = 0.072764 \; \frac{N}{m} = \frac{kg}{s^2} \quad \text{liquid surface tension}$$

$$\mu_L = 0.0010042 \; \frac{kg}{m\,s} \quad \text{liquid viscosity}$$

$$\rho_L = 998.20 \; \frac{kg}{m^3} \quad \text{liquid density}$$

c. Look up the properties of the critical contaminant, benzene, at the minimum water temperature, T,

$$D_L = 8.91 \times 10^{-10} \; \frac{m^2}{s} \quad \text{liquid diffusivity of benzene } at \text{ 20°C (296.13 K)}$$

d. Obtain data from product literature (Table D-6).[*]

Table D-6
Packing Characteristics

$$d_p = 0.0508 \; m \quad \text{nominal diameter}$$

$$a_t = 157 \; \frac{m^2}{m^3} \quad \text{total surface area}$$

$$s_c = 0.033 \; \frac{kg}{s^2} \quad \text{critical surface tension for polyethylene packing}$$

$$c_f = 15 \quad \text{packing factor}$$

e. Liquid mass velocity is as follows.

$$L \;=\; \rho_L \,\frac{Q_L}{A}\left[\frac{\text{kg}}{\text{m}^2\,\text{s}}\right]\; \text{liquid mass velocity at } 0.01388\,\frac{\text{m}^3}{\text{s}}\; \text{with a nominal column diameter of } 0.76\,\text{m}$$

$$=\; 998.19 \times \left(\frac{0.01388}{0.45599}\right)$$

$$=\; 30.38\;\frac{\text{kg}}{\text{m}^2\,\text{s}}$$

f. Calculate the Reynolds Number, N_{Re}.

$$N_{Re} \;=\; \frac{V_L\,\rho_L}{a_t\,\mu_L}\quad \text{(Reynolds Number)}$$

$$V_L \;=\; 0.3043\,\frac{\text{m}}{\text{s}}\quad \text{from Paragraph D-4}d$$

$$\rho_L \;=\; 998.19\,\frac{\text{kg}}{\text{m}^3}$$

$$a_t \;=\; 157\,\frac{\text{m}^2}{\text{m}^3}$$

$$\mu_L \;=\; 0.0010042\,\frac{\text{kg}}{\text{m s}}$$

$$N_{Re} \;=\; \frac{0.3043 \times 998.19}{157 \times 0.0010042}$$

$$=\; 192.7$$

$$N_{Re}^{0.1} \;=\; 1.692$$

g. Calculate the Froude Number, N_{Fr}.

$$N_{Fr} \;=\; \frac{a_t V_L^2}{g_c}\quad \text{(Froude Number)}$$

$$=\; \frac{157 \times (0.3043)^2}{9.807}$$

$$=\; 0.01483$$

$$N_{FR}^{-0.05} \;=\; 1.234$$

h. Calculate the Weber Number, N_{We}.

$$N_{We} = \left(\frac{1}{a_t}\right)\frac{V_L^2\rho_L}{g_c s} \quad \text{(Weber Number)}$$

$$= \left(\frac{1}{157}\right)\frac{(30.39)^2 \times 998.19}{9.807 \times 0.072764}$$

$$= 0.08094$$

$$N_{We}^{0.2} = 0.6048$$

i. Calculate the wetted area of the packing, a_w from the dimensionless relation:

$$\frac{a_w}{a_t} = 1 - \exp\left[-1.45\left(\frac{s_c}{s}\right)^{0.75}\left(N_{Re}^{0.1}N_{Fr}^{-0.05}N_{We}^{0.2}\right)\right]$$

$$N_{Re}^{0.1}N_{Fr}^{-0.05}N_{We}^{0.2} = 1.692 \times 1.234 \times 0.6048$$

$$= 1.263$$

$$\left(\frac{s_c}{s}\right)^{0.75} = \left(\frac{0.033}{0.0728}\right)^{0.75}$$

$$= (0.45352)^{0.75}$$

$$= 0.553$$

j. Calculate the wetted surface area.

$$\frac{a_w}{a_t} = 1 - \exp\left[-1.45(0.553 \times 1.263)\right]$$

$$= 1 - \exp(-1.0125)$$

$$= 1 - 0.3633$$

$$= 63.67\%$$

$$a_t = 157\ \frac{m^2}{m^3}$$

$$a_w = 63.67\%\,(157)$$

$$a_w = 99.96\ \frac{m^2}{m^3}$$

$k.$ Calculate the liquid phase mass transfer coefficient, Onda K_L from the following relationship:

$$K_L \left(\frac{\rho_L}{\mu_L g_c} \right)^{\frac{1}{3}} = 0.0051 \left(\frac{V_L \rho_L}{a_w \mu_L} \right)^{\frac{2}{3}} \left(\frac{\mu_L}{\rho_L D_L} \right)^{-0.5} \left(a_t d_p \right)^{0.4}$$

$$\left(\frac{\rho_L}{\mu_L g_c} \right)^{\frac{1}{3}} = \left[\frac{998.19}{(0.0010042)(9.8066)} \right]^{\frac{1}{3}}$$

$$= (101{,}361)^{1/3}$$

$$= 46.63$$

$$\left(\frac{V_L \rho_L}{a_w \mu_L} \right)^{\frac{2}{3}} = \left(\frac{0.3043 \times 998.19}{99.96 \times 0.0010042} \right)^{\frac{2}{3}}$$

$$= (302.7)^{\frac{2}{3}}$$

$$= 45.08$$

$$\left(\frac{\mu_L}{\rho_L D_L} \right)^{-0.5} = \left[\frac{0.0010042}{(998.19)\left(8.91 \times 10^{-10} \right)} \right]^{-0.5}$$

$$= (1129)^{-0.5}$$

$$= 0.02976$$

$$\left(a_t d_p \right)^{0.4} = \left(157 \times 0.0508 \right)^{0.4}$$

$$= \left(7.9756 \right)^{0.4}$$

$$= 2.2946$$

$$K_L \left(\frac{\rho_L}{\mu_L g_c} \right) = 0.0051 \left(\frac{V_L \rho_L}{a_w \mu_L} \right)^{\frac{2}{3}} \left(\frac{\mu_L}{\rho_L D_L} \right)^{-0.5} \left(a_t d_p \right)$$

$$K_L = \frac{0.0051 \times 45.08 \times 0.02976 \times 2.2946)}{46.63}$$

$$K_L = 0.0003367 \, \frac{m}{s}$$

l. Calculate the gas phase mass transfer coefficient, Onda K_G, using a stripping factor (R) between 2 and 5. Try $R = 2.5$ if air pollution control is required, $R = 4.5$ if it isn't.

$$\frac{K_G}{(a_t D_G)} = 5.23 \left(\frac{G}{a_t \mu_G}\right)^{0.7} \left(\frac{\mu_G}{\rho_G D_G}\right)^{\frac{1}{3}} (a_t d_p)^{-2.0}$$

m. Look up the properties of the gas (air) at the minimum water temperature, T (Table D-7).

Table D-7
Air at 20°C (293.16 K) and 1 atm

$$\mu_G = 1.773 \times 10^{-5} \frac{kg}{m\,s} \quad gas\,viscosity$$

$$\rho_G = 1.2046 \frac{kg}{m^3} \quad gas\,density$$

n. Look up the properties of the critical contaminant, benzene, at the minimum water temperature, T.

$$D_G = 9.37 \times 10^{-6} \frac{m^2}{s} \quad gas\,diffusivity\,(benzene\,in\,air\,at\,20°C,1\,atm)$$

o. Calculate the gas flow rate from the relationship:

$$\frac{Q_{G_{min}}}{Q_L} = \frac{(C_{ai} - C_{ao})}{H'_a C_{ai}}$$

$$= 4.253\,from\,Table\,D\text{-}4$$

$$V_L = 0.03044 \frac{m}{s}$$

$$V_{G_{min}} = 4.2635 \times 0.03044$$

$$= 0.1297 \frac{m}{s}$$

$$R = 3.5$$

$$V_G = R \times V_{G_{min}}$$

$$V_G = 3.5 \times 0.1297$$

$$= 0.4531 \, \frac{m}{s}$$

$$G = V_G \rho_G$$

$$= 0.4531 \frac{m}{s} \times 1.2046 \frac{kg}{m^3}$$

$$= 0.5458 \frac{kg}{s \, m^2}$$

p. See Table D-6 for packing characteristics, a_t and d_p.

$$\left(\frac{G}{a_t \, \mu_G} \right)^{0.7} = \left(\frac{0.5458}{157 \times 1.773 \times 10^{-5}} \right)^{0.7}$$

$$= (196.06)^{0.7} \qquad \text{Gas phase Reynolds number}$$

$$= 40.24$$

$$\left(\frac{\mu_G}{\rho_G \, D_G} \right)^{\frac{1}{3}} = \left(\frac{1.773 \times 10^{-5}}{1.2046 \times 9.37 \times 10^{-6}} \right)^{\frac{1}{3}}$$

$$= (1.571)^{\frac{1}{3}} \qquad \text{Gas phase Schmidt number}$$

$$= 1.162$$

$$\left(a_t d_p \right)^{-2.0} = \left(157 \times 0.0508 \right)^{-2.0}$$

$$= \left(7.976 \right)^{-2.0}$$

$$= 0.01572$$

$$a_t \, D_G = 157 \times 9.37 \times 10^{-6}$$

$$= 0.001471 \, \frac{m}{s}$$

$$\frac{K_G}{\left(a_t D_G\right)} \quad = \quad 5.23 \times 40.24 \times 1.162 \times 0.157$$

$$= \quad 3.846$$

$$K_G \quad = \quad 3.853 \times 0.00147$$

$$= \quad 0.005658 \; \frac{m}{s}$$

q. Calculate the overall mass transfer coefficient, Onda K_{LA}.

$$\frac{1}{K_{LA}} \quad = \quad \frac{1}{H_a \, K_G \, a_w} + \frac{1}{K_L \, a_w}$$

$$= \quad \frac{1}{0.2320 \times 0.005658 \times 99.96} + \frac{1}{0.003367 \times 99.96}$$

$$= \quad 7.622 + 29.71$$

$$= \quad 37.33$$

$$K_{LA} \quad = \quad 0.02679 \; s^{-1}$$

$$HTU \quad = \quad \frac{V_L}{K_{LA}}$$

$$= \quad \frac{0.03044}{0.02679}$$

$$= \quad 1.136 m$$

r. Determine *NTU* for the selected *R.*

$$R \quad = \quad \frac{G}{G_{min}}$$

$$= \quad \frac{H_a}{P_{T_o}} \times \frac{G}{L}$$

$$= \quad 3.5$$

$$NTU \quad = \quad \left(\frac{R}{R-1}\right) \ln \left(\frac{\left[\left(\frac{x_{ai}}{x_{ao}}\right)(R-1)\right]+1}{R} \right)$$

$$NTU = \left(\frac{3.5}{3.5-1}\right)\ln\left[\frac{\left[\left(\frac{750}{10}\right)(3.5-1)\right]+1}{3.5}\right]$$

$$= \left(\frac{3.5}{2.5}\right)\ln\left(\frac{(75\times2.5)+1}{R}\right)$$

$$= 1.4\times\ln\left(\frac{187.5+1}{3.5}\right)$$

$$= 1.4\times\ln\left(\frac{188.5}{3.5}\right)$$

$$= 1.4\times\ln 53.86$$

$$= 1.4\times 3.99$$

$$= 5.88$$

$$Z = NTU\times HTU$$

$$= 5.88\times 3.07$$

$$= 17.13\text{m}$$

$$\frac{A}{W} = \frac{0.4132\ \frac{\text{m}^3}{\text{s}}\ \text{air}}{0.02776\ \frac{\text{m}^3}{\text{s}}\ \text{water}}$$

$$\frac{A}{W} = 14.89$$

s. Calculate the system headlosses, including the packing, the stripper inlet, and the exit losses. Size equipment, including blowers and pumps. Verify that blower discharge pressure is less than the value that would cause flooding.

D-7. Complete the Design.

a. The following drawings are required.

(1) Site plans.

(2) Profiles.

(3) Layout drawings.

(4) Details.

b. Design Analysis

(1) Narrative.

(2) Documentation.

(3) Description.

(4) Calculations.

(5) Computer print out with documentation.

© J. Paul Guyer   2021

5.5.5.5 SUMMARY. The potential and applicability of each NOx reduction technique is summarized in table 11-4.

Technique	Potential NO_x Reduction (%)	Advantages	Disadvantages
Load Reduction	See Figure 11-1	Easily implemented; no additional equipment required; reduced particulate and SO_x emissions.	Reduction in generating capacity; possible reduction in boiler thermal efficiency.
Low Excess Air Firing (LEA)	15 to 40 see Table 11-2	Increased boiler thermal efficiency; possible reduction in particulate emissions may be combined with a load reduction to obtain additional NO_x emission decrease; reduction in high temperature corrosion and ash deposition.	A combustion control system which closely monitors and controls fuel/air rations is required.
Two Stage Combustion			
Coal	30	-	Boiler windboxes must be designed for this application.
Oil	40	-	
Gas	50	-	Furnace corrosion and particulate emissions may increase.
Off-Stoichiometric Combustion			Control of alternate fuel rich/ and fuel lean burners may be a problem during transient load conditions.
Coal	45	-	
Reduced Combustion Air Preheat	10-50	-	Not applicable to coal or oil fired units; reduction in boiler thermal efficiency; increase in exit gas volume and temperature; reduction in boiler load.
Flue Gas Recirculation	20-50	Possible improvement in combustion efficiency and reduction in particulate emissions.	Boiler windbox must be modified to handle the additional gas volume; ductwork, fans and controls required.

Table 11-4

Comparison of NOx reduction techniques

www.ingramcontent.com/pod-product-compliance
Lightning Source LLC
Chambersburg PA
CBHW042040110726
48006CB00002B/243